Alexandre François Augustin Liautard

Animal Castration

Alexandre François Augustin Liautard

Animal Castration

ISBN/EAN: 9783744724678

Printed in Europe, USA, Canada, Australia, Japan

Cover: Foto ©berggeist007 / pixelio.de

More available books at **www.hansebooks.com**

ANIMAL CASTRATION.

BY

A. LIAUTARD, M.D., H.F.R.C.V.S.,

Professor of Anatomy, Operative Surgery and Sanitary Medicine to the American
Veterinary College, New York,
Foreign Corresponding Member of the Société Centrale de Medecine Vétérinaire,
Paris,
Member of the Société Vétérinaire d'Alsace-Lorraine,
Member of the Société Vétérinaire Pratique, Paris,
Member of the U. S. Veterinary Medical Association,
Etc., etc.

With Forty-four Plates embodied in the Text.

v

NEW YORK
WILLIAM R. JENKINS
VETERINARY PUBLISHER AND BOOKSELLER
850 Sixth Avenue
LONDON: BALLIÈRE, TINDALL & COX
1884

TO

HENRY BOULEY,

MEMBER OF THE INSTITUTE OF PARIS,

This little practical work is respectfully dedicated in testimony of the continued esteem and respectful remembrance of one who was fortunate in being numbered among his former students, by

THE AUTHOR.

INTRODUCTION.

In presenting this concise treatise upon castration of the domestic animals, it is not intended to offer new modes of operation, but merely to collect together the various methods in use and leave the reader to appreciate them at their value. It is a work which, it is believed, has not yet been done in English veterinary literature, and on this account it is hoped will prove of interest and use to those engaged in that specialty of veterinary surgery.

In gathering the material, advantage has been taken of several of the most recent works of French and German writers on the subject, and plates have been obtained from the original and excellent wood cuts of Zundel, and Peuch, and Toussaint.

In presenting this volume to the indulgence of veterinarians, it is with pleasure that the author acknowledges and offers his sincere thanks to Dr. Holt for the great assistance he has kindly given in revising the manuscript.

THE AUTHOR.

IMAL CASTRATION.

ANIMAL CASTRATION.

CHAPTER I.

DEFINITION—ITS VARIOUS PURPOSES—AN OLD OPERATION—ITS HISTORY—CASTRATION OF NECESSITY—CASTRATION OF FASHION AND CONVENIENCE—ITS EFFECTS—UPON THE GENERAL ORGANISM—UPON SOME SPECIAL FUNCTION—UPON THE DEVELOPMENT OF THE ANIMAL—AGE AT WHICH IT OUGHT TO BE PERFORMED—SEASON MOST FAVORABLE—PREPARATION OF THE PATIENT—MODES OF RESTRAINT—CASTING—STANDING UP—ANATOMY OF THE PARTS.

OF all the operations pertaining to the domain of Veterinary Surgery, without doubt the practitioner is most frequently called upon to perform—more especially in breeding districts—that of castration, the destruction or removal of the essential organs of generation in our domesticated animals. It is, however, not nearly so often resorted to for purely surgical reasons as for purposes closely related to questions of agricultural and industrial economy, by reason of its effect upon the individuals of the various species of animals subjected to it, in order to improve their value and increase their usefulness to mankind. And that this is its practical effect is no modern discovery. As far back in antiquity as seven centuries preceding the Christian era, it was known and practised upon various animals. Of this we may find ample historical proof in the writings of Roman, Greek, and Oriental authors, where specific mention appears of the various methods employed, including the processes of excision, of crushing and of tearing. Even the castration of females was known to our less remote ancestors, the Danes having in the sixteenth century performed it on sheep, swine, cows, and even mares. The spaying of cows, however, seems to have been forgotten about the beginning of the present century, and it was not until the year 1831 that Thomas Winn, of Natchez (Louisiana), and afterwards Levrat (of Lausanne), brought it to the attention of veterinarians, as a means for the improvement of the milky secretion in cows.

The operation may be considered under two distinct heads. Under the first it is to be considered as one of *necessity*, as when performed with a therapeutic object in view; as, for example, when it constitutes one of the first steps involved in the surgical treatment of strangulated hernia, or of diseased conditions of the testicles or ovaries, such as orchitis, epididymitis, sarcocele, hydrocele, cyst, etc. In the other case it is resorted to for reasons of mere *fashion* and *convenience*, and has for its object the production of such a modification of the general organism as shall increase the adaptedness of the animals subjected to it to the uses to which they are applied; when, of course, the economic becomes the paramount and exclusive reason for thus interfering with the obvious creative purpose. It is the operation as performed under this general heading that we shall now principally consider.

In relation to this latter object it must be borne in mind that the operation is followed by certain peculiar effects, which may either manifest themselves upon the entire organism, or upon some special functions only. In the first instance it is quite evident that the primary and most obvious effect of the mutilation is to be discovered in the character and disposition of the animal, which at once becomes in a double sense an "altered" creature, docile and submissive, and entirely willing to become the obedient and useful servant of his human master. But it is not alone that we find the vicious stallion, the uncontrollable bull, the kicking jackass, the dangerous boar, and even the hysteric mare and cow transformed into the useful gelding, the quiet ox, the patient donkey, and the "fatherly" barrow, the quiet working mare and the productive cow, as the result of the change which the character—the nervous system, in fact—has undergone. Besides this, other marked changes are to be observed of a more distinctly physical character, such as a modification of the entire organism, manifested in the external symmetry, and the expressive physiognomy of the creature, when deprived of its virility.

The animal becomes more quiet, and its general form is modified. If altered at an early age, the skeleton will be arrested in its growth, and the mass of muscles attached to it will participate in the defective development; the head will become elongated, the legs will continue to be lighter, and the body will show a corresponding lack of development. In other words, the male animal will tend to assume the characters of the female, in form and feature, the gelding, indeed, resembling the mare, not only in the *ensemble* of his appearance, but in his voice, which loses the resonance of the stallion's, and his physiognomy, which becomes milder and less expressive; while his neck is lighter and his mane more scanty, with the hairs which compose it more fine and silky.

A like tendency exists in other male animals to acquire a resemblance to the female as an effect of the operation of castration. The altered bull has a weak and feminine voice; his head is narrower and elongated; his horns become lengthened and more curved; he has exchanged his wild and threatening aspect for a mild and gentle visage; his neck also is lighter and his chest narrower; his bony structure is less massive; and he has, besides all the rest, acquired a quality of essential importance to mankind in a dietetic view, that of accumulating fat. This last phenomenon shows us that besides the other changes referred to, there is an important modification of the nutritive forces of the animal, or at least a change in the direction of their action.

When thus deprived of his virile functions the animal ceases, in effect, to exist as one of a species, but maintains an essentially individual life, in which the assimilable nutriment which he absorbs, instead of being in part appropriated to the office of reproduction of his kind, is all devoted to his own individual conservation. In animals not used for draught purposes, or in other labor, when the food received is nearly always in excess of the amount required for the support of the organism, the result follows that the surplus of nutritive substances (found sometimes in great abundance) becomes stored in the connective tissue and intermuscular structure, and that in this way the flesh assumes superior and more nutritious qualities than that of the unaltered animal, while, at the same time, it loses the strong and peculiar odor frequently communicated to it by the presence of the testicular apparatus and secretion in the entire animal.

This property of modification of function is probably still better illustrated in the effect of the operation upon cows, where we shall find not only the power of accumulation of fat increased by castration, but, above all, the milky secretion improved both in quality and quantity, and also in the duration of the flow.

AGE.

The question, "at what age can an entire animal be altered?" admits of a simple answer, to wit, "as soon as the testicles can be easily reached—as soon as they appear outside of the abdominal cavity, and are found in the inguinal canal." But although it can be performed at that epoch, or deferred to any period of after life, it must be remembered that it is easier and less dangerous in young than in older animals, and that with the former it is a simple operation, producing, ordinarily, no noticeable alteration in the other functions, and but rarely followed by accidents.

A period between eighteen months and two years is generally preferred for horses, though, according to some authors, even a much earlier date may be chosen, some English veterinarians being accustomed to operate at as early a date as ten days from birth. It is immaterial, however, at what precise time the operation may be performed, since it is a conceded point that the earlier it is done the better.

SEASON.

When it is possible to choose the season most favorable for the operation, and for securing the best chances of recovery, the spring, or the early stages of the fall, are those to which the operator should give the preference, provided the atmospheric temperature is moderate and not susceptible to sudden variations. It is to be remembered that at some periods of the year, without any known or apparent cause, a tendency appears in wounds to take on gangrenous or septicemic complications which are not so generally observed in the mild weather of spring and early fall. Another essential condition which surgeons will do well to take into consideration is the general health of the subject, as in all cases of surgical interference, any diseased tendency already existing (perhaps latent) in the patient, such as an anæmic condition, a gourmy predisposition, or typhoid susceptibility are likely to give rise to the development of serious and perhaps fatal sequelæ to an operation which, simple as it may be in itself, is nevertheless not without danger, or of possible complications of its own.

PREPARATIONS.

The preparations to which the animal is to be subjected previous to undergoing the operation are the same as those which are required in other cases of surgical manipulation. Some portions of the preparations are, perhaps, of even greater importance, and may not, on any account, be overlooked, when we take into consideration the peculiar position in which the animal must frequently be secured in order effectually to control his movements. Hence, a low diet for twenty-four hours preceding that appointed for the operation, and an empty stomach at the time of castration, with a thorough washing of the sheath, are precautions which no surgeon entitled to the name will overlook or neglect, especially when a soliped is to be subjected to the knife. We shall discuss hereafter the indications in the case of the castration of large females.

RESTRAINT.

Two modes of restraint are employed in securing the animals during the manipulations for the removal of the testicles, one which is applied to all the various methods yet to be described, and the other applicable principally to the method of amputation of the cord by the use of the *écraseur*. In the former, the animal is thrown down and secured with one of his hind legs fixed in a position in which the inguinal region is fully exposed. In the latter he is allowed to remain in a standing posture, and is kept quiet by the application of a twitch upon his upper lip. As the first mode of securing the patient is the safest for all parties engaged in the undertaking, and from the further fact of its applicability in all methods of operating, we shall first consider it somewhat in detail.

By veterinarians who employ the old method of casting with four hobbles, the animal, being properly prepared, is thrown upon whichever side corresponds with the operator's habit of manipulation, whether with the right hand or the left, and the leg opposite to that on which he is lying being released from the hobble, is carried forward upon the corresponding shoulder, as far as it can be safely done. To effect this a loop of rope or platelonge is passed around the coronet, below the fetlock, the free end being carried forward over the dorsal border of the neck, under the neck, towards its anterior border, and is then carried back under the same hind leg, between the hinder extremities and over the hock, from the

FIG. 1.

Condition of the horse in lying posture. Steps to bring one of the hind legs upon the corresponding front one.

posterior border, where an assistant, stationed at the back of the animal, is ready to receive it (Fig. 1). By careful, gradual and steady pulling upon the rope the foot is brought forward upon the external surface of the shoulder, and there secured by two or three turns of the rope around the coronet.

But it often occurs that in this position the inguinal region is not sufficiently exposed, and some of the steps of the operation may thus be rendered difficult, even when the surgeon has taken the precaution to pose the body of the patient and place him partly on his back, by means of bundles of straw pressed under the side upon which he lies. Many operators prefer the use of the double side line, with which, when the animal is thrown, both hind legs are brought forward together, and he is fixed squarely upon his back, and the inguinal region thus brought distinctly into view. The manipulation is thereby made easier for the surgeon, and, it is claimed, safer for the patient. The possibility of danger attending these methods has led many veterinarians in Europe, and in the United States as well, to prefer the operation with the animal in the standing posture. But it is to be considered that the same complications may rise in all instances, with the exception of apprehended injury to the vertebral column, which, though possible, is almost unknown, in fact, at the usual age of the castrated animal. It is to be considered, likewise, that the animal, on his feet, is free to struggle as violently as he wishes, and is thus exposed to the risk of the pulling and laceration of the spermatic cord, and a resulting predisposition to enlargements of that body and the formation of champignons. When it is considered, again, that by the traction of the cord the superior opening of the inguinal canal is necessarily

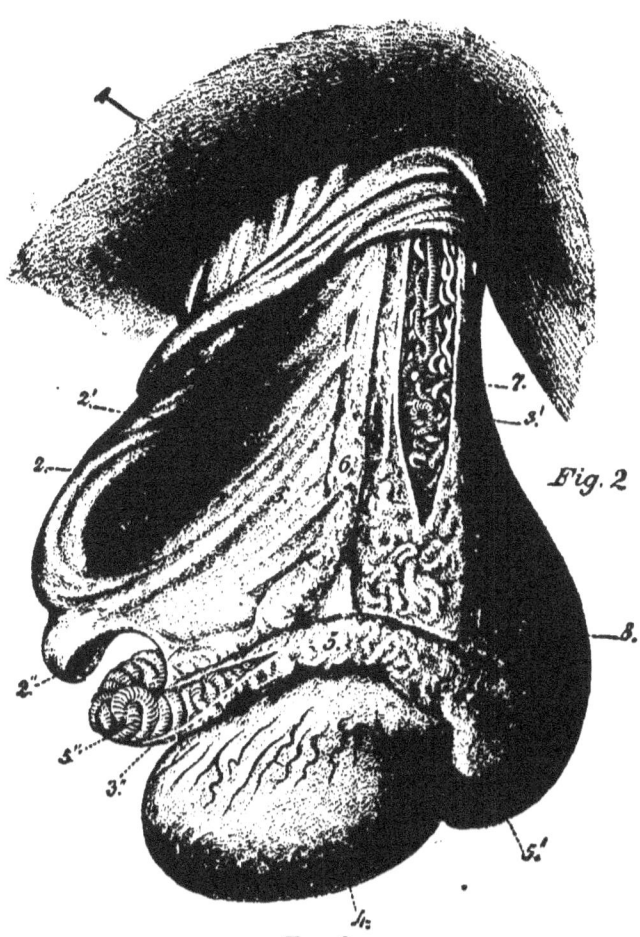

FIG. 2.
1. Testicular envelope. 2. Posterior serous septum. 2'. White muscular fibres of Bouley. 3. Serous membrane—portion of the tunica vaginalis 3'. 3". Visceral layers of the tunica covering the cord and the testicles. 4. The testicle with its peritoneal covering. 5. The epididymis. 5'. Globus major. 5". Globus minor, the tail. 6. Deferent canal. 7. Spermatic blood vessels and nerves.

dilated, and the formation of a hernia of castration is liable to take place, we submit the point to the intelligent judgment, whether, in the presence of these possibilities of extremely dangerous accidents, it does not become the duty of the veterinarian to prefer the mode of securing his patient in the supine position, both in his own behalf and that of his employer.

ANATOMY.

We now pass rapidly in review the anatomical structure of the inguinal region and of the testicular organs. A knowledge of these is of course essential to a proper understanding of the description of the various modes of operation, and of some of their sequelæ, upon which we shall soon enter.

The testicular envelopes, passing from the surface inwards, are represented by the scrotum, the dartos, the cellular coat, the tunica erythroida, formed by the cremaster, and the fibrous and serous or vaginal sac (Fig. 2). The scrotum is a continuation of the skin, and forms a complete bag, common to both testicles, which it contains and covers; the skin being here thin, vascular and nervous, usually black in color, almost hairless, and soft and unctuous to the touch. It is divided into two lateral halves by a *raphœ* or median line. It is very elastic, and easily yields to the efforts of distention, to which it is subjected, and when stretched over the organs it contains, presents a shining aspect, due to the sebaceous secretion which covers it. It easily contracts to its shrunken condition, and may be closely drawn up into the inguinal canal, when it assumes a thickly wrinkled surface.

The second envelope, the dartos, is a prolongation of the tunica abdominalis, and is a yellow, fibrous structure, forming two distinct sacs resting upon each other, and lying on the inside of the scrotum, to which it is intimately adherent.

In the lateral and superior parts the adhesions are looser, and in front it becomes continuous with the suspensory ligament of the sheath, which, like itself, forms a portion of the abdominal tunic. Under the dartos is a layer of very loose cellular tissue, the lamellæ of which are so formed that it may be divided into several superimposed layers. This formation endows the testicle with great mobility in the dartoid sac; and these layers may be easily separated with the finger from the external surface of the fibrous coat beneath, except posteriorly, where it forms a strong band which sometimes requires even the aid of an instrument to divide.

The next envelope is represented by the tunica erythroida which is the cremaster muscle, and from the lumbar region extends itself downwards into the inguinal canal along the outside of the cord, and terminates towards the superior part of the testicle in fibres spreading only over its external face. This muscle, by its deep surface, rests upon the fibrous coat—another envelope of the testicle and of the cord—and to which it is closely adherent. To the powerful contraction of this muscle is due the retraction of the testicle into the depth of the groin, which condition sometimes it is so difficult to overcome in the first stages of the operation.

The fibrous testicular envelope which we have just seen giving attachment to the cremaster, is a thin membranous bag, elongated like the neck of a bottle around the spermatic cord, which it envelopes, and dilated below, in order to enclose the testicle. Lined internally by the serous coat, to which it intimately adheres, this last membrane is a duplicature of the peritoneum, drawn downwards by the testicle when it descends from the abdominal cavity into the inguinal canal. This serous envelope has, therefore, two coats, one lying on the inside of the fibrous tunic, and called the parietal, and that which covers the cord and the testicle and is known as the visceral. These two layers approximate towards the posterior border of the cord, and, as they unite, form a sort of fold, band, or septum which divides into two parts the posterior portion of the vaginal cavity, and becomes a means of solid adhesion between the tail of the epididymis and the bottom of the sac.

The testicles, thus covered by the visceral layer of the serous coat, are suspended at the end of the spermatic cord, and surmounted upon their superior border by the epididymis, the first part of the deferent canal, which is folded upon itself, while at its posterior extremity—the "tail," so called—it continues in a straight course, and conveys the product of the secretion of the testicles into the vesiculæ seminales, lodged in the pelvic cavity.

The spermatic cord is formed anteriorly by the spermatic or great testicular artery, which forms, in that portion, a large number of flexuosities, causing its length greatly to exceed that of the cord to which it belongs. It contains a network of veins, and lymphatic vessels in abundance, which are united to the curves of the artery by a somewhat loose cellular tissue. A large number of nervous branches, given off by the solar plexus, surrounds the whole.

Between the lamellæ of peritoneal structure which forms the posterior septum, and which unites the parietal with the visceral layer, there is found a band of grey muscular fibres—first discovered, I believe, by H. Bouley—which exerts a powerful agency in the retraction of the testicle towards the inguinal ring. Behind this muscle, and situated on the internal face of the septum, are found the deferent canal and the circumvolutions of the small testicular artery.

Having thus considered the essential points of the anatomical structure of these organs, we shall next seek to enforce the importance of their careful study in reference to the intelligent and skilful performance of the important operation which we are discussing.

CHAPTER II.

CLASSIFICATION OF THE METHODS—THREE CLASSES—FIRST, IMMEDIATE AMPUTATION—SECOND, AMPUTATION AFTER APPLICATION OF HEMOSTATIC MEANS UPON THE CORD—THIRD, WITHOUT AMPUTATION, BUT DESTRUCTION OF SECRETING POWER OF THE ORGAN—SIMPLE EXCISION—SCRAPING—TEARING AND TORSION — TORSION — FREE AND LIMITED — ABOVE OR BELOW THE EPIDIDYMIS—LINEAR CRUSHING—FIRING.

THE methods of performing the operation of castration may be variously classified, though in each class a varying number of modes will come under our notice.

The first class will include the operations by which, the envelopes having been cut through, the vaginal sac opened and the testicle exposed, the organ is separated by an immediate section of the cord. A number of different processes are included under this head, among which are those of *scraping*, of *tearing*, of *torsion*, of *linear crushing*, or by the *ecraseur*, and of *firing*, or the actual cautery.

The second class has also for its first or preliminary step, that of the first, viz., the incision of the bags, the opening of the vaginal sac, and the exposure of the testicle. But instead of removing the organ by the division of the cord, we proceed as a second step, to the application of an apparatus designed to operate by producing compression along the length of the cord, and in this are included but two modes of operating, that by the *ligature* and that by the *clamps*.

The third class, according to our category, presents to our view two further operations, both of which are essentially bloodless and dispense with the incision of the bags, consisting of certain peculiar manipulations which insure the destruction of the testicular structure, and consequently of its secreting power. They comprehend the process of the *crushing of the spermatic cord*, and that of *subcutaneous double twisting*—the *bistournage* of the French.

We now enter upon the consideration of each of the separate modes we have thus enumerated.

SIMPLE EXCISION.

This is claimed to be one of the oldest modes of operating, and though to a great extent discarded by practitioners of the present day, still finds its application in the treatment of the smaller animals. With larger patients, however, though still strongly recommended by some practitioners, it is not generally employed on account of the profuse hemorrhage which necessarily follows the amputation of the cord. Still it is conceded that this hemorrhage, as in many cases of the clean, transverse section of arterial blood vessels, will cease spontaneously by the contraction of the vessels on themselves, and the formation of a clot at the divided end, as well as in the surrounding cellular tissue. However, there is a possibility of the continuance of the hemorrhage for some length of time, and the bad effects of excessive depletion are not to be overlooked, especially in an animal whose general constitution has from any cause suffered impairment. If there is any one of the various modes of operating in which the standing position is allowable, this, in our opinion, is the one, the steps of the process being so few and so short, and admitting such simplicity and rapidity. These consist in making a free opening in the bag, reaching with a single stroke of the knife into the vaginal cavity; grasping the testicle and pulling it gently downwards; and cutting the cord right across, from the front backwards, above the epididymis, the cord returning of itself into the vaginal sac—the division being made, of course, on both sides. The animal is then kept quiet in his stall and left alone until the hemorrhage subsides.

SCRAPING.

This operation, which is said to have originated in India, is but a modification of the preceding. Instead, however, of using a sharp edged instrument to divide the cord, the surgeon, on the contrary, employs a dull knife, with which the coats of the artery and portions of the cord are scraped until the separation takes place. They are thus placed in good condition for their temporary closure. This method is probably attended with a diminished amount of hemorrhage, and if carefully performed, it may be entirely absent, the clot closing the artery, and the condition of the lacerated threads of the vessel acting favorably, as well, in preventing it. The manipulations are similar to those accompanying the simple excision, though it is better and more safely effected when the animal is on his back. The testicles being exposed by the incision through the envelopes, the posterior septum of the cord is cut through by a transverse section, and the scraping of the anterior fasciculus of the cord then performed, by a slow movement from above downwards, along a certain extent, in order to effect a solution of continuity by a sort of wearing through the tissues. This operation is slow and requires a careful hand for its execution. But as it may in some cases be followed by severe hemorrhage, it cannot, for that reason, be recommended for large animals, for solipeds especially.

TEARING AND TORSION.

These two modes of operation may, to a great extent, be considered as identical. Indeed, the mode of torsion may be said to have arisen principally as a modification of that of tearing, which is the older. In tearing, the cord was subjected to a certain amount of torsion *by the hand*, and then torn apart at a given point in its length; while in the process of simple torsion, as properly performed, we obtain a division of the cord by twisting it *with instruments*, which enables the operator to effect the separation at a definitely determined spot. Tearing differs, then, from torsion only in the fact that after giving several twists to the cord in order to gather its fibres into a more compact mass, and to diminish the resistance of the more superficial layers, it is divided in its continuity by a violent traction upon its fibres in the direction of its length.

In this process, especially applied to ruminants, the testicle being exposed, the operator secures the cord firmly with the thumb and index finger of one hand, to prevent the traction from taking effect too far upwards when being made by the other hand, after the cord has been twisted a few times on itself.

TORSION.

In this method of castration the cord is twisted with sufficient force to cause it to break of itself at the point of the greatest violence. Its design is to accomplish the removal of the testicle without dragging or excessive traction upon the cord, and thus to avoid the hemorrhage following the torsion of the spermatic artery, as a mode of hemostasis sufficient to prevent the flow of blood attendant upon the rupture of the cord. The operation may be performed either above or below the epididymis, or may consist simply in the torsion of the artery alone. At first the hands only were called upon to act in the manipulations, and the operation was from this cause known as *free torsion*, until about fifty years ago, when instruments were introduced into general practice, and gave rise to the plan of *limited torsion*.

FREE TORSION.

Free torsion, or that in which the hands alone are employed in the operation, may be performed, as before stated, either above or below the epididymis.

Above the Epididymis.—The first is one of the oldest modes of castration known; one which must have been practiced contemporaneously with the use of clamps, or in the first age of surgery. The first steps of the operation required for the exposure of the testicles are the same as have already been detailed. When this is accomplished the steps of torsion and rupture are then performed in the following manner: The operator, grasping the testicle, carefully draws out the spermatic cord, and with a pointed bistoury makes a transverse incision, above the tail of the epididymis, through the posterior septum of the cord, involving what we know as the white muscle of Bouley, the efferent canal, and the small testicular artery. He then seizes the anterior fasciculus of the cord between the thumb and index finger of the left hand, squeezing it as tightly as possible, and having with the other hand secured the cord at a short distance below the point where the left hand has already been placed, performs the torsion by a rotatory movement given to the testicle itself, the motion having for its result the twisting and tearing of the cord when long enough continued to overcome the tenacity of its fibres. Fifteen or twenty turns of the organ will usually be found sufficient to effect the rupture. A considerable degree of strength in the fingers is required in this movement, and for this reason the torsion may take effect further up than may be desired, and beyond the point designed, which may result in an unnecessary amount of irritation and injury. When the torsion has gone so far that the rupture of the cord has been effected, the stump is released, and retracts in the inguinal canal to a certain height limited by the presence of the posterior septum, which holds it in place, and to a great extent prevents its return through the superior orifice of the inguinal canal.

Below the Epididymis.—This, the fourth step of the operation, consists in the separation of the testicle from the epididymis and the torsion of one upon the other. The testicle being exposed, the operator, taking hold of its appendix, the epididymis, with the left hand, and of the gland with the right, their cellular serous attachment is divided by the thumbs from the posterior to the anterior extremity, from the tail to the head of the twisted efferent canal. If this cannot be done with the hand, the convex bistoury must be called into action. This accomplished, the head of the epididymis is firmly secured with the fingers of the left hand, and the right hand, left free, gives to the testicle the number of rotatory motions necessary to separate it from its excretory canal— that is, from eight to ten. When the testicle is thus severed, the stump of the end, with the epididymis, is pushed back into the vaginal sac, where it is confined by the application of a suture upon the middle of the edge of the scrotal wound.

LIMITED TORSION.

Limited Torsion Above the Epididymis.—As we have seen, this is the operation by which the division of the spermatic cord is effected by torsion made upon a given point in its length, and limited by the use of special instruments.

We have already called attention to the difficulty

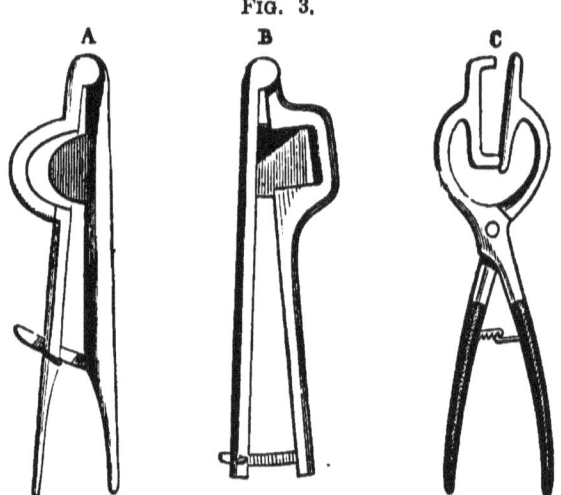

FIXING FORCEPS FOR TORSION.
A.—Renault and Delafond pattern.
B.—Perier.
C.—Reynal.

of the operation of free torsion, which requires a great deal of strength, and which, besides, may be accompanied by a serious inflammatory condition of the parts, through rough manipulations of the cord. It is for this reason that this mode of procedure must have been reserved for small animals, as, if performed upon the larger kinds, it can only be by men whose muscular force is sufficient to enable them to overcome and bring into subjection the struggling subjects of their operations.

It was in 1883 that two French veterinarians, Renault and Delafond, of the Alfort school, introduced the use of instruments in the operation, as an improvement upon the manual methods and their effects on the sequelæ, though it is said to have been

FIG. 4.

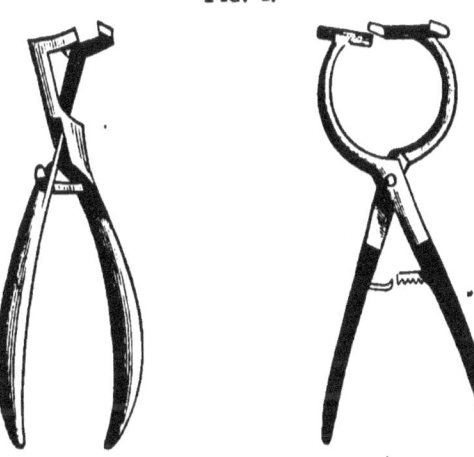

MOVING FORCEPS FOR CASTRATION BY TORSION.
Renault and Delafond pattern.　　Reynal pattern.

already practiced in Germany as far back as the last century. The instruments employed are two forceps of peculiar construction, and which were more or less modified, one of which (*fixing* forceps, Fig. 3,) is to be applied upon a fixed point of the cord, where it is suffered to remain, and the other (*moving* forceps, Fig. 4,) is employed to accomplish the rotation of the testicle and the lower end of the cord. Those of Renault and Delafond or of Reynal are now in general use. Those of Beaufils (Fig. 5) are, we believe, too complicated for general use.

Modus Operandi.—In the first step of the operation, the ordinary manipulations of the division of the envelopes, the opening of the sac and exposure of the testicle being accomplished, and the envelopes being carefully pushed upwards, the torsion and excision of the cord are effected in the following manner: The entire cord is embraced by the fixing forceps (see Fig. 6), or only its anterior fasciculus if the posterior septum has been cut, as in the process by free torsion above the epididymis. An assistant, seizing it from before backwards between its open branches and strongly closing them, holds it firmly, without pulling upon the cord. The operator then grasps the cord with the moving forceps above the testicle, and a little below the point held by the assistant, leaving a small space between the instruments, and closing his own tightly, begins the movement of torsion, which he directs from left to right. For this he sometimes requires both hands, one of them keeping the instrument in place, while the other continues

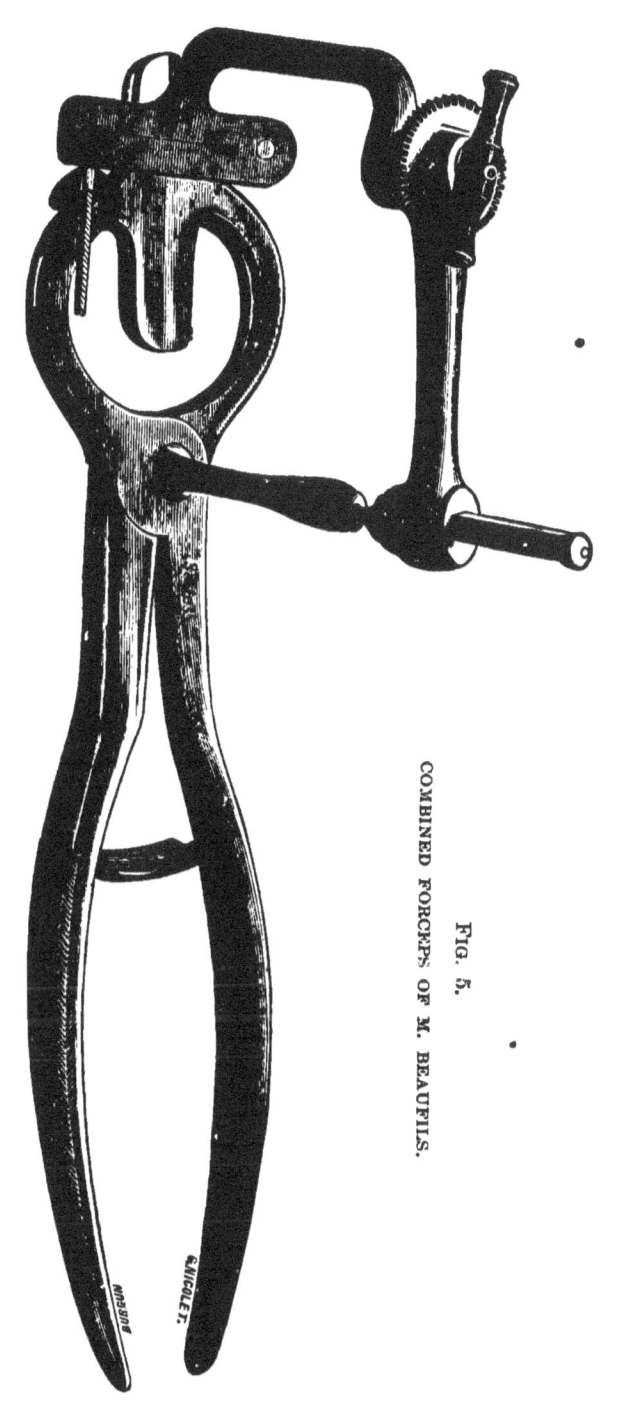

Fig. 5.
COMBINED FORCEPS OF M. BEAUFILS.

the rotation as described. Ten or fifteen turns of the forceps are usually sufficient to complete the rupture

Fig. 6. OPERATION OF LIMITED TORSION.

of the cord, the artery, owing to its facility of elongation, being the last part to give way. The testicle then separates, being held in the branches of the moving forceps; the fixing forceps are removed, and the cord is drawn upwards into the vaginal sac. It is important in this operation to caution the assistant against drawing on the cord during the struggles of the animal, consequent upon the pain caused by the first application of the instruments, and the pressure upon the parts when held between their inflexible iron jaws; but on the contrary, to maintain it as closely as possible against the inguinal region.

This process of castration is one of the most rapid of all the forms of operating. The only hemorrhage likely to occur is merely that of the small testicular artery, if it should happen to be divided when the torsion is confined to the front portion of the cord.

Below the Epididymis.—This process differs from the preceding only in the fact that instead of holding the cord between the fingers, it is held by the fixed forceps, the use of the moving instrument being rendered unnecessary by the slightness of the adhesion of the seminal gland to its appendix.

LINEAR CRUSHING.

The originator of the use of that peculiar instrument, the ecraseur (Fig. 7), so valuable an adjunct in the operation of castration, is Mr. H. Bouley, who brought it into use at a date as early as the year 1857. It is not, therefore, an American invention, as has sometimes been claimed. The function of this instrument is to effect the division of living parts without hemorrhage. The original ecraseur of Chassaignac has received many modifications, all of which, however, operate upon the same principle

FIG. 7.

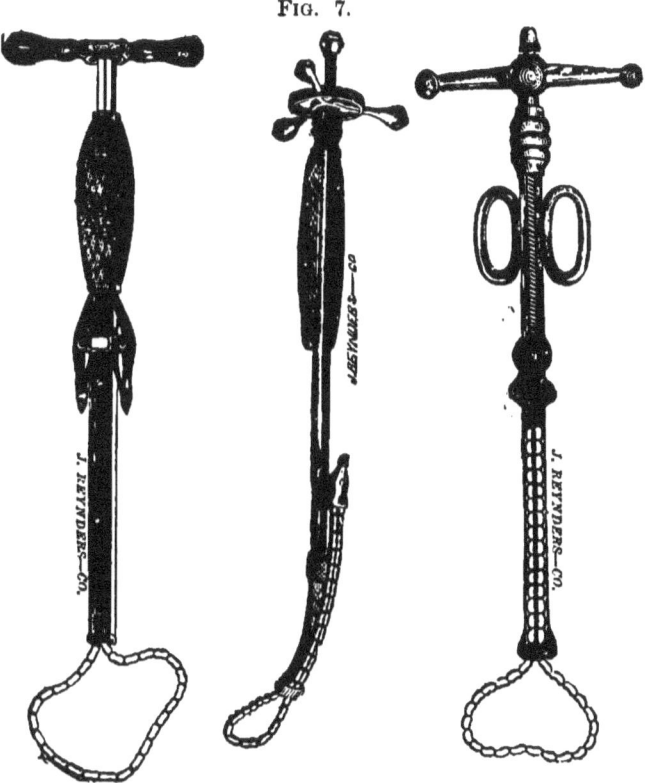

VARIOUS KINDS OF ECRASEURS.

The essential design of all is to produce a general constriction of the blood vessels, by which their internal and middle coats being first divided, may contract within the cavity of the vessel in such a manner as to close their cavity and form a sort of stopper to the artery, while the external cellular covering, the last to undergo division, is so stretched, under the action of the instrument, and so closely adapts itself by its ends, that insufflation through the free ends of the vessels fails to remove the closing arrangement of the two coats first divided.

Modus Operandi.—The operation is comparatively a very simple one. The testicle being exposed, as in all the other methods, the chain of the instrument is so placed around the cord that the pressure takes place upon the greater mass of tissue, in order that it may continue the longer; which being done, the lever of the instrument is brought into action, and the constriction caused by the chain slowly kept up until the definite division of the tissue is accomplished. The essential condition of success in the operation, having in view the desired hemostatic effect, is to *act slowly*. According to Prof. Bouley, an interval of several seconds should be suffered to elapse after each rotation of the wheel which moves the chain. If the tissues are divided too rapidly, the section of the artery is apt to be too clean, and a hemorrhage is likely to be the result. This objection, however, though made by one of the highest authorities in veterinary surgery, does not seem to be justified by the results obtained by American operators, most of whom both recommend and practice its execution as rapidly as possible; and according to their own statements, a serious hemorrhage is seldom encountered. The fact that it has been observed in any case, however, confirms the wisdom of the recommendation of Prof. Bouley, and as most of our American *confrères* prefer the operation with the animal in the standing posture, the reason of their neglect of the prudent and more truly surgical process can be readily appreciated. To avoid the hemorrhage Dr. House invented a clamp which he applied upon the cord previous to the amputation with the ecraseur (Fig. 8). This mode of castration is not very exten-

FIG. 8

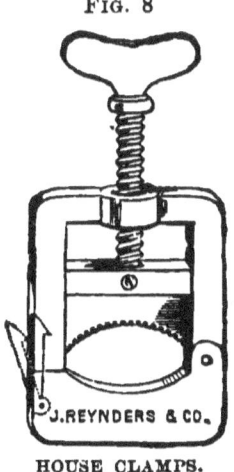

HOUSE CLAMPS.

sively practiced by European practitioners. The reason of this is probably to be looked for in the essential necessity of safety which so protracts its performance.

FIRING.

This mode of castration consists in the application to the cut end of the testicular cord—previously divided with the bistoury, or by the cautery—as a means of hemostasis, of an iron heated to a

FIG. 9.

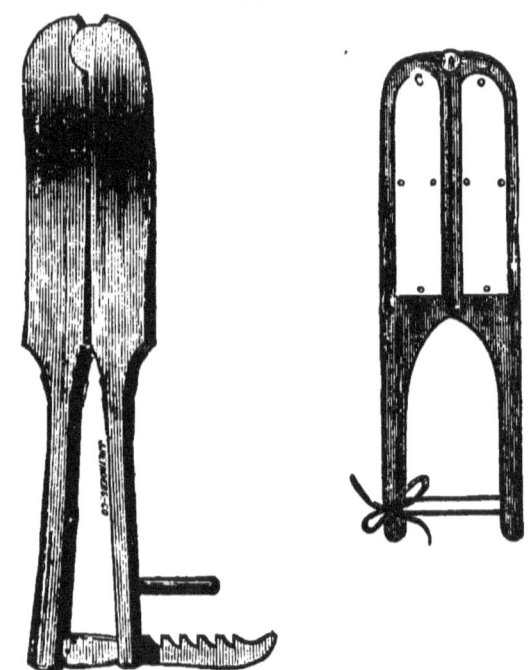

SINGLE FORCEPS FOR CASTRATION BY FIRING.

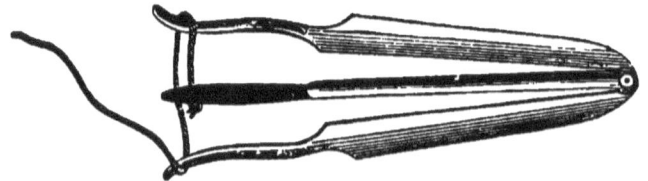

DOUBLE FORCEPS FOR CASTRATION BY FIRING.

white heat—the actual cautery This is claimed to be one of the oldest modes of operating, Vegetius and Absyrtus describing it as a common process of castration. It is much in favor in England, and in some parts of Germany, though less practiced in some other parts of Europe. The instruments essentially needed for the operation of castration by firing are two; the first, a peculiar forceps for holding the cord and securing it while the application of the cautery is being made; and the second, the iron or cautery itself.

These forceps, or nippers, are either single or double (Fig. 9), and may be made either of wood or of iron, and more or less modified in form, according to the fancies of the different operators. But they all work on the same principle, and effect the same object. With the single forceps but one cord can be treated at a time, but with the double instrument both cords may be secured at once, and may be divided and cauterized at one step. In this way the possibility of disturbing the eschar caused by the cauterization of one cord while manipulating the second, is quite obviated.

When the testicles, either or both, have been exposed, the mass of the cord is fixed between the jaws of the forceps, from before backwards, at about one inch above the tail of the epididymis, and firmly secured. This may be effected either by tying it tightly with a string wound about the handles, or by means of a spring crank with which some instruments are furnished. The testicle is then amputated, either by a stroke of the bistoury, or with the sharp edge of the cautery carried across and at a right angle with the direction of the cord. This done, the operator applies the broad portion of the iron over the entire surface of the stump of the spermatic cord, and cauterizes (or sears) the part thoroughly. It must be remembered that to insure the safety of the cauterization, the iron must be very hot. Otherwise, when it is removed, if it has cooled off, it may adhere to the carbonized surface, and the scab formed at the end of the blood vessel may accompany the instrument. The application of pulverized rosin to the end of the cord, previous to the cauterization, is recommended by some practitioners.

A very proper precaution, and one on no account to be omitted, is the protection of the surrounding parts from the radiating heat by covering them with wet cloths.

When the operation is completed, the forceps should be opened with great care, in order to ascertain whether all hemorrhage has ceased, and the cord may be allowed to retract. If any oozing of blood appears at the point of the operation, the cauterization must be repeated at the point indicated.

A free application of cold water, in the form of a *douche*, after the operation, will contribute to the formation of a clot in the cauterized artery.

CHAPTER III.

METHODS OF THE SECOND CLASS — CLAMPS—COVERED AND UNCOVERED OPERATION—INSTRUMENTS—FOUR VARIOUS STEPS—OBJECTION TO THIS METHOD—TIME TO REMOVE THE CLAMPS—LIGATURE—OF THE CORD AND ITS ENVELOPES—OF THE CORD ONLY—OF THE SPERMATIC ARTERY—OF THE EFFERENT CANAL—SUBCUTANEOUS OPERATION.

Having completed the consideration of the various methods included in the first class, we propose next to examine those entering into the second, which embrace those in which certain means of pressure are applied and suffered to remain upon the cord previous to the amputation of the testicle. These are two in number, and consist of the process known as that of the clamps, and that which involves the use of the ligature.

THE METHOD BY THE CLAMPS.

This is an ancient mode of operating, having been transmitted to us through many ages. It has received the sanction of long practice, and, if not absolutely superior to all others, is possessed of qualities and advantages which all who have employed it will freely acknowledge.

It is performed in two ways. One is the process of the *covered*, the other of the *uncovered* testicle. The covered operation is that in which only a portion of the testicular envelopes are divided, the scrotum and the dartos, the gland being left covered with the other envelopes. On the other hand, in the uncovered operation, all the enveloping membranes are divided, and the testicle is made to protrude outside of the vaginal sac. The first three steps of this mode of operating are understood to have been performed in the methods which we have already considered as generally preliminary in all cases, in order to obtain access to the cord.

The instruments necessary to operate in this case are a very sharp convex bistoury, a pair of clamps, some strong twine, a castrating forceps and a pair of scissors. The clamps are wooden or metallic pincers, formed to embrace the cord and to be applied firmly upon it, in order to hold it securely, and to confine the artery tightly enough to prevent the occurrence of hemorrhage. The form most ordinarily used, and probably most convenient, is made of wood, and consists of two semi-cylindrical pieces (Fig. 10) joined at one end and resting together by a flat and sometimes grooved surface, and measuring about six inches in length. The material is a light but strong wood. They are rounded at the extremities, in order to avoid chafing the soft tissues. A groove at each end is de-

ANIMAL CASTRATION. 41

signed to receive the twine, which is part of the appliance. Before being used they are tightly tied together at one end, in such a manner that they

FIG. 10.

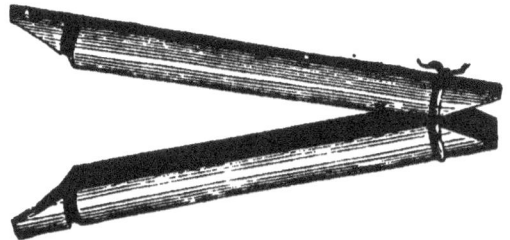

ORDINARY CLAMP.

INTERNAL FACE OF CLAMPS.

spring open if closed. This point is one of the first importance. It requires considerable exercise of strength to place them tightly enough on the cord they are to hold, but this firm juxtaposition, so obtained, facilitates their removal, when that is required. The groove which they carry on their flat surface is sometimes filled with some merely lubricating greasy substance, usually simple ointment, fresh lard, butter, or cream, though some veterinarians use a caustic paste. This last mode of proceeding is strongly opposed by some authorities, as likely to induce unnecessary inflammatory action, through the formation of a scab, which may require for its removal a process of sloughing, which may in some cases give rise to serious complications. As I have stated, the clamps are not always made of wood, and very many alterations and improvements, so called,

FIG. 11.
VARIOUS SHAPES OF CLAMPS.

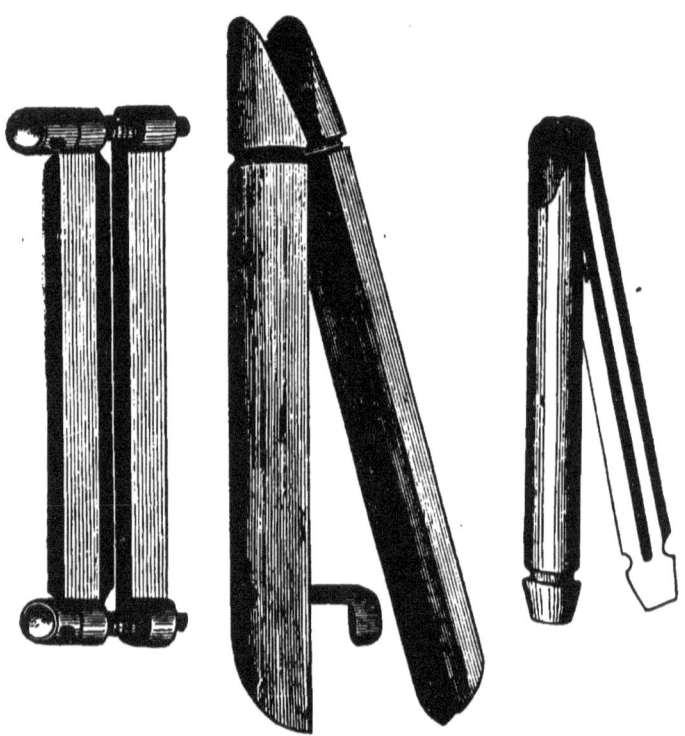

Screwed Clamp.   Spring Clamp of Brandt.   Hinge Clamp.

have been from time to time brought forward. In Fig. 11 a few of these clamps are presented. The oldest form is the simplest, and possesses the further advantage of being always easy to be obtained, while the more complicated contrivances are not always easy of access.

The castrating forceps (Fig. 12) are used for bring-

FIG. 12.

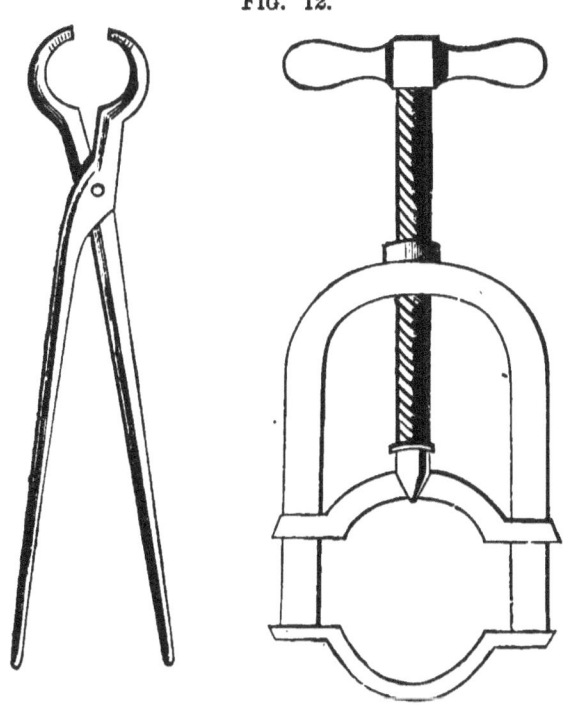

FORCEPS TO HOLD THE CLAMP TOGETHER   VISE FOR THE SAME.

ing the clamps together while they hold the spermatic cord between their branches. There are several kinds of these, but in default of obtaining them readily, the operator may find an eligible substitute in the blacksmith's nippers or the gasfitter's tongs. We have used this latter for a good many years, and have found it very well adapted to the purpose required, by the presence of its set of double curved and grooved jaws. The twine which it is necessary to use to keep the clamps closed when they have been brought into perfect contact with the castrating forceps, must be soft and strong. A piece of fishing line, previously waxed, will answer the purpose very well. In order to facilitate the traction which may be necessary to keep the clamps in place, it is a good precaution to attach the ends of the twine to small wooden handles to protect the hands from cutting by the string. It is well, also, to prepare a reserve of clamps and twine against accidents from breakage or the mislaying of these articles.

COVERED OPERATION.

Modus Operandi.—The animal being thrown on either side, as already described, and kept as nearly as possible on his back by bundles of straw packed under him on the lower side, and the right hind leg secured in its proper position, and the instruments placed within easy reach, the surgeon proceeds with the fourfold steps of the operation, consisting first, in the prehension of the left testicle, or lowest in position; second, the incision through the envelopes; third, the enucleation of the testicle; and fourth, the application and constriction of the clamps.

First step.—The operation must always begin with

the prehension of the gland which corresponds with the side—the lower—upon which the animal is lying. This obviates any danger of interference by any little hemorrhage which might occur, and so facilitates the application of the clamps. Then, placing himself toward the back of the patient, the operator reaches over and grasps the lower testicle with both hands, bringing it downwards in such a manner as to stretch the scrotum over its surface. This manipulation is not always of easy performance, the contraction of the cremaster muscle being sometimes so powerful that the gland successfully resists all the operator's efforts of traction. It is sometimes necessary to divert the attention of the animal, in order to facilitate this part of the process, by pricking him with a pin on the lips or about the anus, the effect of the new sensation being such that his opposition is withdrawn, and the contraction ceasing, he suffers passively the traction of the envelopes over the organ. Or, the same advantage may be obtained by the inhalation of a little ether or chloroform. Then grasping the cord with the left hand and bringing the organ well forward, the surgeon proceeds to the

Second step, or that of the incision of the envelope. Holding the sharp convex bistoury in his right hand, he takes, with the thumb, a *point d'appui* upon the prominent organ, and carries it carefully over the surface of the scrotum in a direction parallel with the median raphè (described in the first chapter), and following the great curvature of the testicle, and being careful with the first movement of the instrument to divide only the scrotal skin and the dartos, until the most superficial layers of the cellular tissue of the third testicular envelope are reached. The skin and the dartos being divided, the edges of the wound separate, and the testicle, still pressed downwards and outwards with the left hand, protrudes more or less, still included, as it is, within its fibrous covering. A careful dissection, with a few light strokes of the bistoury, or laceration with the thumb nail of the hand, now suffices for the separation of the fibrous envelope from its external covering, an entire separation of both of which can thus be easily obtained by pressing the most external layer upwards through the laceration of the cellular coat which unites them.

Third step.—The operator now relieves himself of his instrument—not, we may venture to suggest, by placing it between his teeth, as some careless surgeons are apt to do, but by handing it to an assistant —and, changing his position, places himself in front of the inguinal region, and facing it. He then proceeds to the enucleation of the testicle, by separating the adhesion which exists between the internal face of the dartos and the external surface of the cremaster muscle and of the fibrous tunic. The separation being completed, and the scrotum and dartos being carefully pushed upwards, the patient is now ready for the last step of the operation.

Fourth step.—The testicle, well enucleated from its superficial envelopes, but still covered by the fibrous coat, and the vaginal sac still remaining intact, the operator, facing, as before, the inguinal region, proceeds to the application of the clamps. The cutaneous covering and the dartos being pushed well upwards, the clamp is placed upon the cord above the epididymis, from before backwards, the assistant, armed with the castrating forceps, taking both of its branches between the jaws of that instrument, carefully bringing them together, and closing them as tightly as possible. The instant of the pressure of the clamp upon the cord is marked by very severe pain, and the suffering animal is excited to powerful struggling. It is important that the assistant should be aware of this, and he should be forewarned to refrain from pulling on the cord, and reminded, in order to avoid injury from this accident, to keep the clamps and the forceps steadily in contact with the inguinal canal. It is probably with a view to the avoidance of this possible injury that the use of a peculiarly constructed vise or forceps has been recommended. The forceps being in place, and tightly confining the branches of the clamps, well adjusted, the operator now applies the twine, and after taking several turns around the grooves of the free ends of the clamps, secures it carefully with a double knot.

The operation is then repeated on the right or uppermost testicle in the same manner, and with the same precautions.

UNCOVERED OPERATION.

The four steps of this operation are the same as those of the previous method, the first requiring the same manipulations and observing the same order, but the second involving some variations. In this the same careful dissection is dispensed with, and one free incision suffices, including all the various envelopes, in order to expose the testicle freely and at once. The incision is made with one free stroke of the bistoury extending from the posterior to the anterior extremity of the testicle, and dividing at once scrotum, dartos, and the fibrous and serous coats. Though this is to be done without hesitation, it is by no means necessary to adopt the practice of some operators, who not only divide the envelopes, but even make a large incision in the testicular structure itself, inflicting thus an unnecessary amount of pain from which the animal might, with a little care, have been spared.

When the surgeon reaches the third step of the operation, and seizes the testicle with the right hand, in order to draw it downward and outside of the vaginal sac, he may encounter great resistance to his traction, from the powerful opposition of the white muscular tissue running along the posterior septum of the cord. He must then slowly and steadily draw the testicle down, and at a given moment, with a single stroke with a sharp pointed bistoury, divide the serous band of the posterior septum, cutting at once the muscular fibres, the efferent canal and the small testicular artery. This being effected, the resistance will terminate, and the testicle may be drawn down without further difficulty. The division of the septum is not always resorted to. Still, the verdict of experience is strongly in favor of the measure. The application of the clamps (Fig. 13) is

FIG. 13.

CASTRATION WITH UNCOVERED TESTICLE.

effected in the same manner as in the covered operation, but in this instance the clamp is placed higher on the cord. For this reason the assistant must be especially careful during the struggles of the patient when the clamps are tightened, the danger of inguinal hernia at this point being too serious to be overlooked. The clamps being in place, and properly secured, the testicles are either left in place and allowed to slough away, or are amputated a short distance below the clamps, as the case may be. The parts being carefully washed out, the animal is allowed to rise, and is returned to his stall.

An objection frequently urged against this mode of operation is that it requires a second visit of the surgeon when the time has arrived for the removal of the clamps. Estimating this objection at its proper value, we consider that it is more than balanced by the advantages attendant upon this special mode of castration, and while we fully appreciate the difficulty and inconvenience to which the surgeon may be subjected by this second visit, we cannot approve of its omission, either from a surgical point of view or in that of the interests of the employers, in whose behalf all care and responsibility should be exercised, until the patient is at least enjoying a fair prospect of recovery.

The question now arises, at what time can the clamp be removed with safety? It must be understood that there may sometimes be peculiar surgical conditions under which their removal is contra-indicated, and when they must be allowed to slough off without further interference on our part. But even in ordinary cases and under favorable circumstances, this time appears to vary. By some they are removed after thirty-six hours, while others allow them to remain for a period of four or five days. Taking a fair average, we are of opinion that it may be safely done on about the third day, and that at that period the closing of the artery is sufficiently assured to remove all further pressure.

If the clamps have been secured with twine, and especially if they were properly prepared previously to their application, the process of removal is a very simple one. The assistant, raising one of the patient's hind legs, the operator places himself directly behind the animal, and bending down, with a sharp sage-knife, cuts the twine where it has secured the posterior ends of the clamp. If it retains the springiness it ought to have possessed at the time of its original application, the branches readily spring open, and it falls to the ground. If this does not occur, or if they should be held by adhesions with some dried parts of the cords which have been pressed between the branches of the clamps, they must be carefully separated by moving from below upwards, when they will easily become detached. But this last manipulation must be very carefully performed, if we would escape a hemorrhage which might require serious measures to control. When clamps of another make are used, the process of removal will vary according to existing peculiarities in the construction of the instrument. The clamp having been removed from one side, the separation from the other will, of course, be managed in a similar manner.

THE LIGATURE.

This method of castration consists in the application of a circular ligature upon the entire cord, or a portion of it, for the purpose of completely closing it, with the various parts entering into its formation. It was in practice so long ago as 1734. The operation is divided into several varieties, viz., that of the cord with its envelopes; that of the cord only, either by the covered or uncovered method; that of the spermatic artery alone; that of the efferent canal; and that by the subcutaneous process.

The ligature used in these various modes of operation is formed of waxed silk; sometimes of strong twine, as fishing line, for example; or, as more recently introduced in surgery in the removal of living growths and tumors, an elastic cord.

Ligation of the cord and its envelopes.—This process is principally used upon small animals, although, since the elastic cord has been brought into use, a few attempts have been made to make it applicable to the larger kinds. The experiments, however, have been as yet so few, and the results so unsatisfactory, and in so many cases fatal, that it can scarcely be recommended, except for small subjects. The application of this is very simple. It consists, after securing the patient, in bringing the testicles as far down into the scrotum as may be thought needful, and after applying the ligature two or three times around the cord, a short distance above them (Fig. 14), slowly and steadily tightening it until a sufficient amount of force has been employed to close the calibre of the blood vessel and cut off the circulation from the parts situated below the point of ligation. This mode of operating has, in our hands, proved very successful in small animals, and when the elastic ligature has been used. Mortification has taken place in a few days, the testicles slowly detaching themselves at the point of ligature, and when falling off leaving but a very small superficial, cutaneous scab, and healing in a short time.

Ligature of the cord only; covered operation.—The first three steps of the operation having been accomplished, and the testicle enucleated, the ligature is

ANIMAL CASTRATION. 53

placed around the cord, still covered by its fibrous envelopes and the cremaster muscle. A piece of twine or an elastic ligature may be employed for

FIG. 14.

CASTRATION BY LIGATION OF THE CORD AND ENVELOPES.

this purpose. In this operation the testicles are allowed to remain not less than twenty-four hours, before amputation is performed, in order that if the pressure has been insufficient, and the parts should fail to exhibit symptoms of loss of vitality after that time, another ligature may be applied.

Uncovered operation.—The only variation between this method and the one last considered is found in the fact that in this, the testicle and cord being exposed as in the process of castration with the clamp, the ligature is applied either on the cord as a whole, or only on its anterior fasciculus. In this case the testicle is amputated immediately after the application of the ligature. But as there is a possibility of the slipping off of the ligature, great care must be taken lest the amputation be performed too near the point where the constriction is made. And again, as there is a possibility of the truncated cord being drawn too far up, even up into the abdominal cavity, it becomes a precaution of prudence, as recommended by Mr. Bouley, to leave a sufficient length of the ligature hanging outside of the scrotal wound, and even to secure it on the edges of the skin.

Ligation of the spermatic artery.—This is a mode of castration which, if we are not mistaken, was held in high estimation by certain practitioners in the city of Boston. It consists simply in the application of a ligature of silk to the spermatic artery. The cord being exposed, and the posterior septum being divided, a curved needle armed with the ligature is made to pass around the whole mass of the anterior fasciculus, and the entire vascular cord is surrounded by the ligature and firmly tightened. The fact of the various and irregular flexuosities peculiar to the spermatic artery, with both ascending and descending portions, explains the necessity of including the entire arterial mass under the ligature, since, if only the simple cord of the artery were ligated, it might be an ascending portion only, and the amputation of the testicle might be followed by a troublesome hemorrhage from one of the descending loops.

Ligation of the efferent canal and the subcutaneous ligation of the cord.—These two modes of operating have not yet yielded sufficient evidence in the form of satisfactory results to be entitled to more than passing mention at the present time. We may say further, moreover, that among all the methods of castration by ligation, none of them have been subjected to a sufficient amount of practical test to be accepted as a process which will justify a strong recommendation or unqualified approval.

CHAPTER IV.

THIRD METHOD OF CASTRATION—CRUSHING OF THE TESTICULAR CORD—DOUBLE SUBCUTANEOUS TORSION—BISTOURNAGE—FOUR STEPS OF THE OPERATION—SYMPTOMS FOLLOWING—CASTRATION OF CRYPTORCHIDS — INGUINAL CRYPTORCHIDY — ABDOMINAL CRYPTORCHIDY—MODUS OPERANDI—FIVE STEPS IN THE OPERATION—EFFECTS FOLLOWING THE OPERATION OF CASTRATION—MODE OF CICATRIZATION—HYGIENIC AND SUBSEQUENT ATTENTIONS OR AFTER CARES.

As I have before stated, the third method of castration embraces the processes in which the testicular envelopes are left intact, while it is the gland or cord which is submitted to the peculiar manipulations by which their structure, and therefore the secreting powers of the testicle, are so essentially modified. In treating of this method, two special operations present themselves for our consideration, to wit: the *crushing of the testicular support*, and the *double subcutaneous twisting*, or *bistournage* of the French. These are employed principally in the castration of ruminants, though efforts have been made to apply the latter in the case of solipeds. The modes of operating which we have already described are, however, also applicable to the ruminants.

CRUSHING OF THE TESTICULAR CORD.

This consists in crushing the spermatic cord with a hammer, the vessel continuing, meanwhile, to be covered with its envelopes. It was first described in the year 1826, and is most commonly practised in some French districts. The instruments used are two cylindrical pieces of wood, each about one yard in length and two inches in diameter; and a hammer or mallet formed of hard and heavy wood. The animal being properly secured in the standing position, the testicles being drawn well down into the bottom of the envelopes, the sticks are placed, one behind and one in front of the cord, close to the upper extremity of the gland. When in that position they are moved in such a manner that instead of remaining, one in front of the other, one becomes so superimposed upon the other that the spermatic cord becomes twisted in the form of the letter S (Fig. 15). While held together in this position by an assistant, the operator, placing himself in front of one side of the hind quarter, with repeated blows of the hammer or mallet, crushes the cord at the point where it rests upon the wood which occupies the inferior position, of course guaging the force and frequency of the blows by the effect observed, until the crushing of the organ is satisfactorily accomplished. As a measure of caution, it will be well, upon the completion of the process of crushing, to surround the cord with a ligature moderately tightened, in order to guard against the drawing up of the cord into the

FIG. 15.

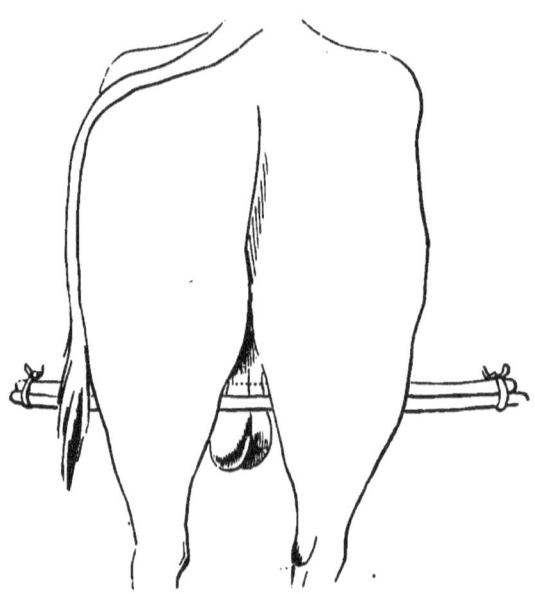

CASTRATION BY CRUSHING OF THE CORD.

inguinal canal, an accident not likely to occur, however, if the operation has been well performed.

DOUBLE SUBCUTANEOUS TORSION; BISTOURNAGE.

In this mode of operating, principally in vogue in the southern parts of France, the position of the testicle is so changed that its lower extremity is made to take the place of the upper, the cord is subjected to a certain degree of torsion, and then the testicle is restored to its normal position, to undergo a process of atrophy which destroys its power of secretion by a physiological action. The great length of the cord and the greater laxity of the cellular tissue situated between the dartos and the fibrous coat, render this operation much easier in the ruminants than in the solipeds. Simple in its manipulations, although still involving a certain degree of dexterity, and followed by comparatively no symptoms of reactive fever, the only instrument necessary for its performance is a piece of cord, twine, or rubber, sufficiently strong to secure the testicular envelopes when the gland has been subjected to the double displacement, and the cord to the torsion it has undergone. In this operation, no special preparation being demanded, the animal is usually treated on his feet.

The late Mr. Serres, of the veterinary school of Toulouse, divides the operation into four steps, viz., *first*, the softening of the bags and separation of the dartos from the fibrous tissue; *second*, the displacement (dislocation) of the testicle; *third*, the torsion of the cord; and *fourth*, the pushing up of the testicles into the inguinal region, with the application of the ligature to keep them in place.

The first step is the most difficult for the surgeon as well as the most painful to the horse, though the contrary is the fact where the subject is an ox. The operator, stationing himself behind the animal, grasps the testicles with both hands (Fig. 16) and quickly draws them down into the scrotum. Hold-

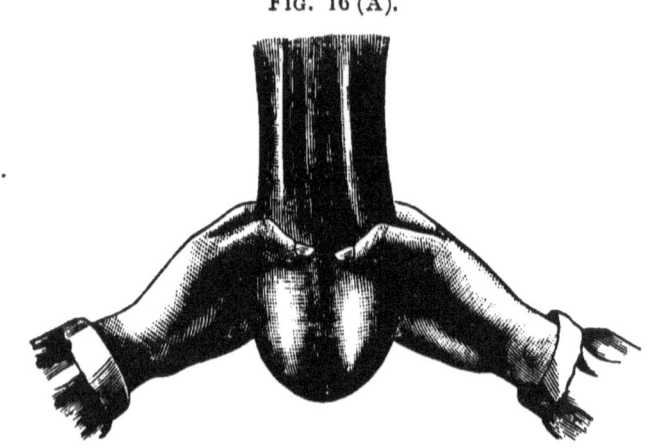

FIG. 16 (A).

DOUBLE SUBCUTANEOUS TORSION IN CATTLE.
Softening the bags—first position of the hands.

ing them there with the right hand, with the left he raises the scrotum by the lower part, firmly pulling upon it downwards and slightly from before backwards. The testicles are then moved upward and downward in the sac, carrying with them the fibrous covering. During this time a peculiar crackling sound is heard, which is caused by the tearing apart of the fibres of the cellular tissue lying between the dartos and the fibrous coat. This laceration is sometimes difficult to effect, especially in aged animals, in which case the up and down motion of the testicles will require a greater number of repetitions before the adhesions are torn.

ANIMAL CASTRATION. 61

The second step consists in the displacement or dislocation of the testicle, which is accomplished in

FIG. 16 (B).

DOUBLE SUBCUTANEOUS TORSION IN CATTLE.

Softening of the bags—second position of the hands.

the manner following: The testicles being pushed well upwards in the vaginal sac, one of them, the left, for example, is drawn well downwards with the left hand, which grasps the cord above the epididymis

(Fig. 17), the thumb resting on the back of the cord, and the remaining fingers in front of it, while the right hand, placed in pronation, pinches the inferior

Fig. 17.

DOUBLE SUBCUTANEOUS TORSION IN CATTLE.
Position of the left hand at the beginning of the second step.

part of the scrotum. Maintaining these dispositions, the testicles are displaced by the simultaneous action of both hands, the left pushing the cord from above downwards and from before backwards, in such a manner as to depress as much as possible the superior extremity of the gland, while with the fingers of the right hand, resting by their dorsal face against the posterior part of the testicle (Fig. 18), the inferior extremity of that organ is pushed upwards. Without losing hold of the envelope, the movement of this hand gives way to these opposite and simultaneous pressures, that of the left hand tending to lower the head of the testicle, and that of the right elevating its tail, and the gland is being flexed upon the cord from which it is suspended, backwards and upwards. At the moment when the testicle forms an acute angle with the cord, the thumb of the left hand, rest-

FIG. 18.

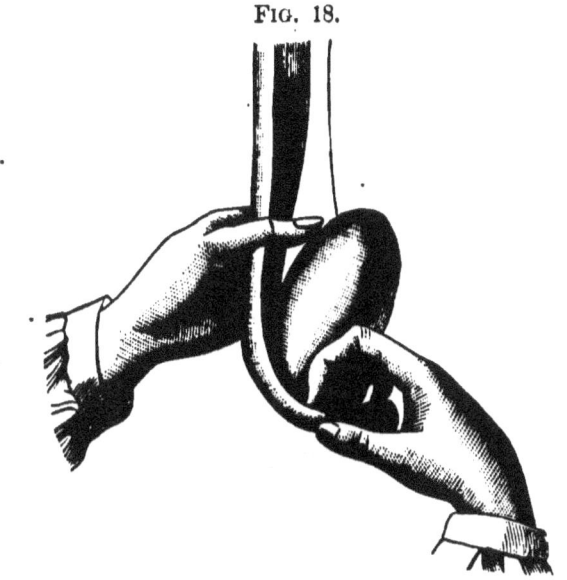

DOUBLE SUBCUTANEOUS TORSION IN CATTLE.
Second step.

ing upon the cord, comes into action to aid in the displacement by making a *point d' appui* upon the inferior extremity of the organ, which now occupies the superior position, in such a manner that the spermatic gland is placed parallel with the cord. The manipulations are completed by pushing the testicles upwards towards the inguinal ring, to break up whatever adhesions of cellular tissue may remain. This second step of the operation being completed, the two organs are found to be so placed that they are parallel one with the other, the testicle being posterior to the cord.

The third step, or that of the torsion of the cord, now presents itself to our notice. To effect this, the

FIG. 19.

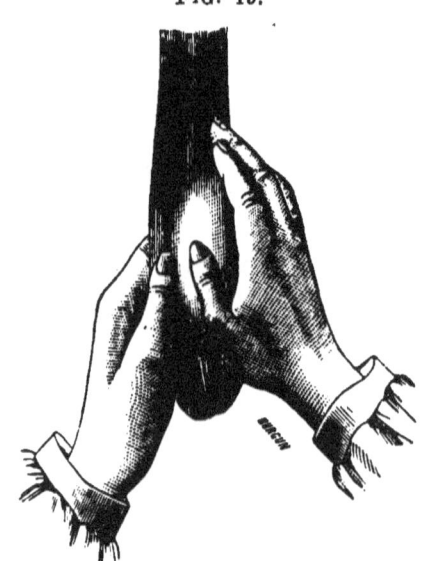

DOUBLE SUBCUTANEOUS TORSION IN CATTLE.

Third step. Position of the hands when the torsion is about being made.

testicle must be firmly held at the bottom of the envelopes (Fig. 19), the left hand placed forward upon the cord, and the right behind and upon the testicle. The operator then gives to the organ a twist with the right hand by a motion of rotation from left to right and from without inwards, while with the other he draws upon the cord in the opposite direction. The result of this manipulation is to give to the gland half a turn around the cord (Fig. 20), which

FIG. 20.

DOUBLE SUBCUTANEOUS TORSION IN CATTLE.

Third step. Position of the hands during the torsion.

thus becomes displaced and takes a posterior position. By a change in the action of the hands, but a repetition of the same movement, the right hand now acting on the cord, while the left is applied to the testicle, the remaining portion of the motion of rotation is performed, and a complete torsion of the spermatic support accomplished. By repeating this action, of course as many turns of the cord as may be thought

66 ANIMAL CASTRATION.

necessary, can be secured, two, however, being generally found sufficient, although, in a few instances as many as four or five may be required—never more than that. The length of the cord is the principal controlling circumstance. When these several steps have been completed with one testicle, their repetition is, of course, in order with the other. And when both have been treated, the consummation of the operation is called for by entering upon the fourth step, or that of the application of the ligature.

To accomplish this both testicles are firmly seized

FIG. 21.

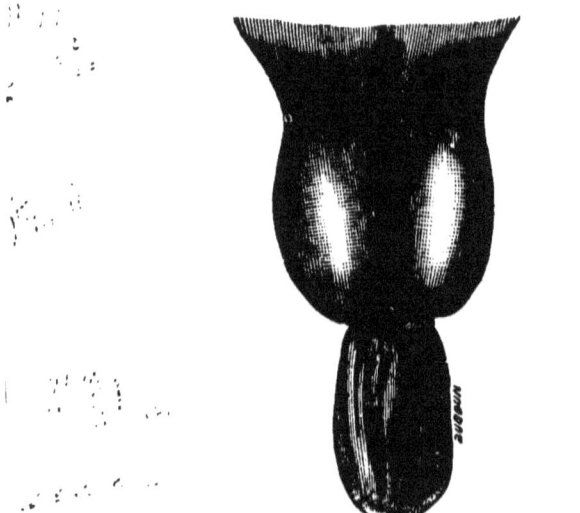

DOUBLE SUBCUTANEOUS TORSION IN CATTLE.

Position of the testicles and ligature en masse of the bags when the operation is finished.

with both hands, and pressed upwards as far as possible against the inguinal ring (Fig. 21). It is necessary to be very careful to ascertain that they rest on the same level, in order to be secure against the possibility of untwisting. The ligature is, then applied by passing three or four turns of it around the scrotal envelopes, immediately below the testicles, with not more than a sufficient degree of tightness to assure it against slipping off.

The symptoms which succeed the operation are not commonly of a very serious nature, and subside within a period of time varying from two to six hours. Following the operation an inflammatory swelling takes place in the bags, and after one or two days assumes large dimensions. The ligature can now be removed, and the animal left to himself, without further treatment, the testicles undergoing a slow process of atrophy readily recognized by their appearance and the position they always thereafter occupy in the vaginal sac.

CASTRATION OF CRIPTORCHIDS.

The abnormal development of animals in which the testicles have failed to make their appearance by descending through the inguinal canal into the bags, is quite commonly met with in solipeds, the animal being then known by the designation of ridglings or originals. The position assumed by the organ in relation to its normal situation being so altered that it may be found either partly engaged in the inguinal canal (Fig. 22), or only remaining close to its superior opening (Fig. 23), is one of these inequalities, constituting what is called inguinal criptorchidy; another being when it remains floating in or adherent to some parts of the abdominal cavity—a condition known as abdominal criptorchidy (Fig. 24). As this condition has usually a peculiar effect on the temper of the animal so affected, often rendering him unfit for general use, it necessitates, on that account, the act of castration, with some changes in the manipulations described for the operation upon animals exempt from such an infirmity. In these cases the operation presents more difficulties, and is of a more serious character than the former, demanding on the part of the operator all the skill and knowledge which can be acquired from its frequent performance and extensive study. That the operation is one which is largely performed on the Continent there is no doubt, and many European operators have made for themselves an extensive reputation in connection with it. Among these the name of Professor Degives, of the Brussels school, merits mention.

But there is probably, on this Continent at least, no better accredited authority, in this branch of surgery, than a gentleman, a layman, of Illinois, known very widely as FARMER MILES, who has for many years not only sustained an eminent repute in his specialty as a gelder, but I believe, has devoted a large share of study specially to the castration of ridglings. He has not only traversed large portions

FIG. 22.

TESTICLE ENGAGED IN THE INGUINAL RING.

B—Testicle. C—Gubernaculum testis. D—Inguinal ring.

FIG. 23.

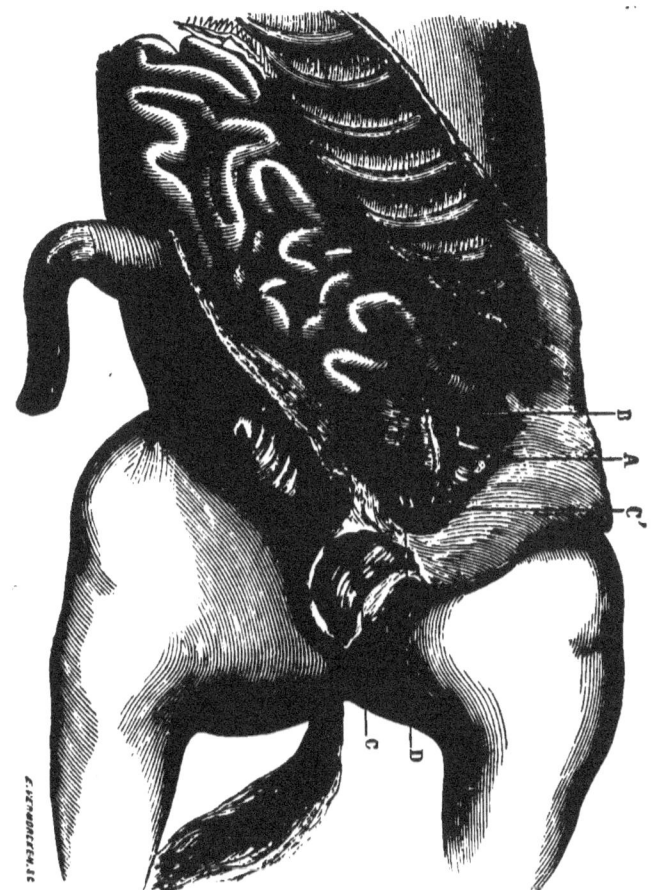

TESTICLE CLOSE TO THE RING.

C'—Internal portion of the gubernaculum testis.
C—Its external portion.
B—Testicle.
D—Inguinal ring.

Fig. 24.

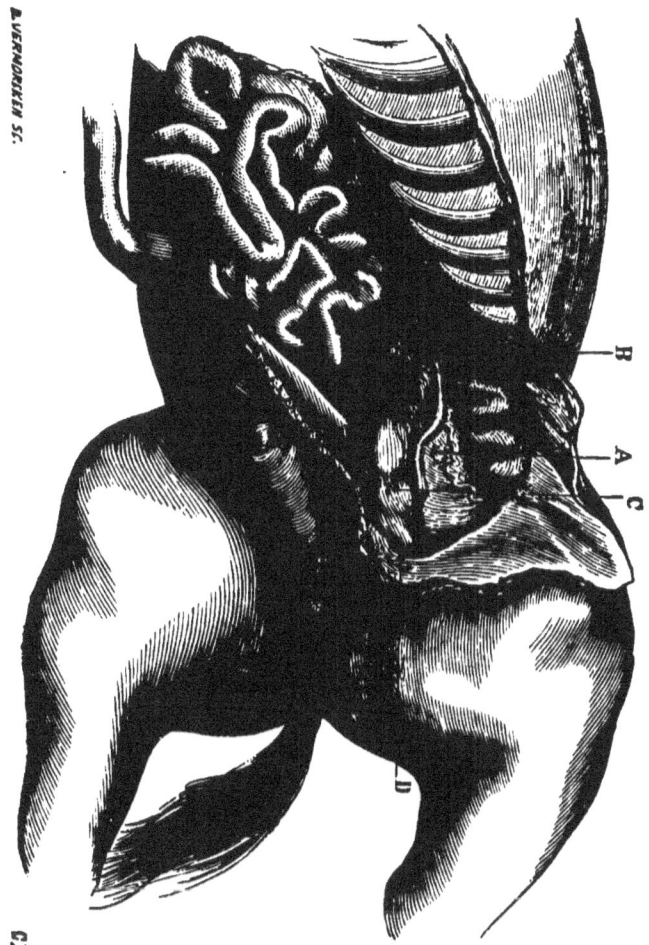

TESTICLE FLOATING IN THE ABDOMEN.

A—Peritoneal ligament attached to the lumbar region.
B—Testicle.
C—Gubernaculum testis.
D—Inguinal ring.

of the United States in the practice of this interesting branch of veterinary surgery, but has likewise achieved much renown and appreciative criticism from foreign sources, having travelled extensively in various European countries, and earned much honor from those who have watched his methods, and received ocular proof of his dexterity and success within the sphere of his special field of usefulness.

The method of procedure, which has in his hands proved so exceptionally successful, though no doubt essentially original with him, is still, we believe, based upon the same principles which govern the operation as we find it described and illustrated in the works of the classical writers who have given their attention to the subject.

We now turn to the consideration of the *modus operandi*, as observed in the two forms of cryptorchidy, the inguinal and the abdominal.

INGUINAL CRYPTORCHIDY.

The preliminary steps in this case are the same as those which are necessary in the castration of animals under normal conditions. The instruments required are a convex bistoury, one or two clamps, or a ligature, and an actual cautery, or the ecraseur most commonly in use. To these is sometimes added a pair of long forceps of peculiar construction, with jaws terminating in two spoon-shaped extremities, designed to grasp the testicle when placed high in the inguinal canal, or if only partly engaged in the ring. Prof. Degives divides the operation into five steps.

First; the incision of the scrotum and the dartos.— The operator, with or without the aid of an assistant, makes a straight longitudinal incision upon the scrotum at the place where the testicle is nominally situated, carefully dividing, also, the yellow fibrous layer which represents the dartos, being especially careful at this point to avoid the large venous branches which abound in the region involved. A sort of hooked bistoury is, we understand, preferred by some veterinarians for this incision, on the score of the additional safety secured by the use of an instrument of that form.

Second step; exposure of the external inguinal ring. —To accomplish this the loose cellular tissue which lies under the dartos is torn and divided by the fingers until the ring is felt. The

Third step is the dissection of the vaginal sheath.— The sheath being situated at varying depths, the dissection is effected by carefully introducing the hand into the inguinal canal, and separating it as much as possible by passing the fingers around its external surface.

Fourth step; opening the sheath.—The opening is made lengthwise, and of sufficient width to allow of the passage of the testicle. When this organ is situated high up in the ring, it is frequently difficult to grasp it and keep it sufficiently steady in position to permit the free use of the bistoury. The sheath being opened and all the testicular envelopes divided, we complete the operation by perfecting the

Fifth stage, or the removal of the testicle.—There are

two ways of accomplishing this, viz. : the direct and the indirect division of the cord. In the former case the amputation is effected either by the process of cauterization, by limited torsion, or with the ecraseur. In the latter the testicle is removed either by a ligature or by the process of the clamps. The process by the ecraseur is at once that which is most generally preferred and the easiest of application.

ABDOMINAL CRYPTORCHIDY.

In this severe form of the trouble under consideration, the various steps of the operation demand careful study. Indeed, so common, so serious, and so frequently fatal are the complications which the surgeon may expect to encounter, that many operators habitually discourage the interference with this peculiar violation of normal conditions.

The first two steps of the operation are similar to those which belong to castration in inguinal cryptorchidy. Following on we have for the

Third step, the perforation of the inguinal canal, or the establishment, by the operator, of an artificial communication from without, with the abdominal cavity within. To effect this the surgeon introduces his hand, with the fingers united in the form of a cone, into the external inguinal ring, and carefully forces them upward towards the external angle of the ilium, resting them upon the crural arch. He soon reaches the closed superior inguinal ring, feeling only the peritoneal membrane, where it is readily torn. Then tearing it sufficiently to permit the passage of the entire hand, or as large a portion of it as is necessary, he has reached the.

Fourth step, or the seizure and removal of the testicle.—The hand, or three fingers, are then passed into the abdominal cavity, in order to feel for the organ or its appendages, until the location is determined, whether of testicle, epididymis, vas deferens, or blood vessels. These are usually found floating not far from the torn opening of the peritoneum. But if not so readily discovered, the hand must be carried above the neck of the bladder, towards the end of the deferent canal, which must be followed until the epididymis or testicle is found. It is then carefully brought outwards by a slow and steady traction upon the testicle itself, or upon a portion of the epididymis, or even upon the extremity of some of the testicular blood vessels.

Fifth step.—The removal of the organ is always much more safely effected with the ecraseur than by other means. The operation is completed by the application of a suture upon the external wound, in order to guard effectually against the possibility of ventral hernia occurring subsequently.

Abdominal cryptorchids are sometimes treated by removal through the flank—an operation intrinsically more dangerous, as well as less promising of success than that in the inguinal region.

EFFECTS FOLLOWING THE OPERATION OF CASTRATION.

These will vary more or less in extent and severity,

according to the method employed in its performance, and in any case they may be considered in two divisions; as primary or immediate, and secondary or consecutive.

Amongst the first phenomena most commonly observed is, of course, a manifestation of pain, characterized by symptoms of colic, exhibited by the animal in a more or less marked degree, being the result of the unavoidable irritation arising from the manipulations practised upon the organs of generation, whose nerves rise from the sympathetic as well as from the cerebro-spinal nervous system; and from the pain excited in the spermatic cord by the pressure of the clamps, for example. These colicky pains, which are more severe under the bloodless method than in those of the other mode, usually subside after the first hour following the operation, and as a rule require but little treatment more than that of the walking exercise. This sort of pain having subsided, the only further trouble likely to be noticed is the local trouble resulting from the lesion to which the testicular region has been subjected. Resulting from this local lesion, as well as from the rough manipulations attending the various steps of the different procedures, a peculiar stiffness will be observed in the motion of the animal. This may be referred either to the local pain proper, to the dragging to which the cord has been subjected, or to the presence of the clamps, which, resting closely in the groin, necessarily more or less impede the action of locomotion.

Hemorrhage may also occur immediately after the operation, either while the patient is still on the ground or as soon as he regains his feet. This may be due either to the solution of continuity at the edges of the wound of the envelopes, or may proceed from the small testicular or the spermatic artery. The first two causes of hemorrhage need not engage our attention, usually ceasing spontaneously, and never being attended with serious inconvenience. It is not so, however, in the case of hemorrhage proceeding from the spermatic blood vessel proper, occurring after those methods of operating which dispense with the closing of the artery by artificial appliances, as is done with the clamp or the ligature, or which may be observed in castration by torsion, cauterization, the use of the ecraseur, or especially by the process of simple excision. Though not necessarily fatal, the hemorrhage in these instances may require prompt and effectual interference by the surgeon for its suppression.

It is not rare for castrated animals to become more or less tympanitic, a condition which may be due, more or less, to the introduction of atmospheric air into the abdominal cavity during the performance of the operation. This condition of things is usually remedied by the unaided action of natural causes.

The secondary effects also vary according to the manipulations of the method which they follow. The development of reactive fever is an event which in many cases requires close watching, and while it is true that many castrated horses will manifest no subsequent illness, even to the extent of a slight elevation of temperature, others, on the contrary, show unmistakable signs of a general inflammatory condition and this is the more marked and definite as the condition of the wound has been left in a more or less complicated state. The presence of the ligature or of a portion of the cord which has yet to complete the sloughing action, following the method by cauterization and by the clamps, are sufficient to encourage the inflammatory tendency.

MODES OF CICATRIZATION.

The cicatrization of the wound of castration takes place in two ways. While the upper part heals by adhesive inflammation at and above the point where the amputation has been performed, it is below that point in a process of cicatrization by the second intention, the parts filling up by the development of granulations, and being accused by an abundant suppurative process. The first fact observed is that the parts become more or less swollen. The swelling is at first limited to the edges of the wound, but increases and spreads to the scrotum, then to the sheath, or even extends forwards and backwards to the perineal region. A flow of serosity will be observed almost immediately following the operation, at first thin and yellowish, but will, before the second or third day, become thicker and more purulent in character, so progressing that after that period it will become a laudable, creamy pus, in evidence of the process going forward towards the establishment of sound and healthy cicatrization. This cicatrization will proceed until the healing is complete—that is, for a period varying between thirty and forty days—the swelling slowly subsiding from the moment when the suppuration becomes established.

THE HYGIENE AND THE SUBSEQUENT ATTENTION.

The moment the patient has risen from his bed and has been thoroughly cleansed from the blood that has soiled his legs, it becomes necessary, if the clamps have been used, to apply the necessary means to prevent their removal by the action of the tail. This is done by braiding the hairs shortly, and sometimes tying it up on the side. Even when this is not necessary, from the clamps not having been used, it is better to have the tail tightened up short, in order, when the suppurative process is established, the more easily to preserve the cleanliness of that part of the body. It is recommended by some veterinarians, also, as soon as the animal is on his feet, to have him thoroughly rubbed and dried, lest, as is not uncommon, he should have perspired excessively during the operation. He may be warmly blanketed if he has been accustomed to a covering, or in any case, placed in a quiet stall and tied up. If quiet and unexcited, and exhibiting no immediate ill consequences of the operation, he may, after an interval, be allowed to go loose in a box stall. If there are any manifestations of pain, or colicky symptoms, walking exercise may be given. Quietness, protection from changes of the weather, moderate diet, varying according to his condition, are included in the only general instructions that can be given.

The wound simply requires to be kept clean. Washing with cool water and soap when the discharge is well established, will fulfil this indication. The closing of the edges of the wound is to be carefully prevented by the introduction of the finger between them, care being taken to avoid the laceration of any points where union has already taken place in the upper part of the wound.

It is not an unusual thing to find even these simple measures of caution overlooked by gelders, some of them even recommending that the animal should, immediately after the operation, be violently exercised—even put in harness and made to draw a wagon. It is true that a little and gentle exercise may be beneficial, with a view to the removal of the soreness and pain of the newly castrated animal; it must be admitted even that Professor Bouley recommends slow exercise to be carried to the extent of fatiguing the animal. But when we take into consideration how seriously some animals, at least, are affected by the operation, and the serious complications which may follow it—even laying aside the humanitarian view of the question—we must necessarily conclude that such directions and such a practice is in violation of all the laws of true surgery, and even if justified by the strongest statistics, is condemned if confronted by a single fatal case.

CHAPTER V.

COMPLICATIONS AND TREATMENT OF THE OPERATION—COLICS—TEARING OF THE CLAMPS—HEMORRHAGE—SWELLING OF THE SCROTAL REGION—GANGRENE—ABSCESSES—CHAMPIGNON—EXTRA SCROTAL—INTRA SCROTAL—INTRA ABDOMINAL—VARIOUS TREATMENTS—FISTULA OF THE SCROTUM—INGUINAL HERNIA—PERITONITIS—TETANUS—AMAUROSIS—COMPARATIVE VIEW OF THE VARIOUS MODES OF CASTRATION.

COMPLICATIONS AND THEIR TREATMENT.

THOUGH the operation of castration is comparatively simple in its various methods and is generally successful in its results, still it is not entirely free from accidents or complications. Indeed, among those likely to meet our notice, there are some of quite a serious character, which will develop themselves independently of the skill and care with which the operation may have been performed or whatsoever attention may have been bestowed upon the patient. Among these may be enumerated *colics, hemorrhage, swelling* of the scrotum, *gangrene, abscesses, champignon, fistula, hernia, peritonitis, tetanus,* and *amaurosis.*

COLICS.

This, we have already seen, usually appears a short time after the completion of the operation, the suffering animal becoming uneasy, restless in his stall, pawing the ground with his fore feet, and sometimes lying down and rolling. As I have before stated, these symptoms, as a rule, are of short duration, and subside without other treatment than a little walking exercise. It is rarely the case that they fail to yield to an anodyne, or a dose of chloral may be demanded before the symptoms are subdued.

TEARING OF THE CLAMPS.

When this accident occurs it is commonly attributable to the omission of a careless operator to secure the tail of the animal in such a manner as to prevent its interference with those impliments by its entanglement, and tearing them from the end of the cord, as a consequence. The result of this is the appearance of a hemorrhage from the spermatic artery, which can only be controlled by either a reapplication of the clamps to the end of the cord—if it can be thus secured—or by other means, which will be considered when we reach the subject of bleeding in general as connected with other causes.

HEMORRHAGE

May be primary or secondary. In the first instance it occurs in consequence of the insufficiency of the means of hemostasis applied to the end of the cord, as in the case of the operation by simple excision, by cauterization, the too rapid crushing of the cord, torsion, or the accident before referred to—when the clamps have been torn off and the cord lacerated about the point of their application.

Secondary hemorrhage manifests itself after a longer interval following the operation. It may occur, for example, after the removal of the clamps, or when, during their removal, the mortified end of the cord is too much interfered with by the sharp end of the instrument used in cutting the string which confines them together, or from too forcibly pulling upon the cord itself; and in some instances without any assignable cause other than a diseased condition of the coats of the artery. This secondary hemorrhage is usually, by reason of the inflammatory condition of the blood vessels, of more threatening aspect and more difficult to control than the primary variety. The treatment indicated varies. When caused by the tearing of the clamps, or at the time of their removal, it may be checked by the reapplication of the instrument. But if the cord is retracted within the inguinal canal and .cannot be reached, and if it is already adherent to the surrounding tissues, by granulations recently formed, the checking of the flow may be very difficult. In many cases the application of cold water, either in the shape of the cold douche over the part, or iced sponges, may prove sufficient. But in other cases the cavity of the wound must be packed with balls of oakum, wet either with water alone or any styptic agent, such as a solution of perchloride of iron, the whole being kept in place by a suspensory bandage, or if necessary, a few points of suture.

These measures may be put in practice while the animal is on his feet; but if they fail in their effect, the surgeon must at once proceed to cast his patient and ligate the artery—an operation of delicate execution, and not always easy to perform, by reason of the deep seated position of the vessel. The use of the actual cautery has also been recommended, but even when successful there are many objections to this, one of which is the complication such an operation may bring on by the introduction into the wound of a scab which must necessitate for its expulsion a serious amount of inflammatory action. As a rule, however, the operation of packing is all that is required, the oakum being left undisturbed for twenty-four or even thirty-six hours. Its removal must be undertaken with great caution.

SWELLING OF THE SCROTAL REGION.

This, as we have seen, is an almost necessary consequence of the operation, the swelling making its appearance a short time after the alteration is accomplished. It usually first affects the parts immediately around the edges of the wound, and spreads forwards and upwards in such a manner that the entire scrotum and sheath become the seat of it. It is somewhat warm, tense, and slightly painful. If there is no increase beyond these limits, there is no occasion for alarm, as by exercise, fomentations, and scarifications, with the administration of diuretics, it ordinarily subsides. But if it continues to increase, and extends upwards and backwards, involving the inside of the thighs and the perineum, loses its character of heat and soreness, to become cold and painless, crepitating under pressure, we must prepare to encounter the most severe of all complications, that of gangrene, requiring the most prompt and vigorous treatment, as we shall presently see. It may also happen that even while retaining the characteristcs of healthy œdema, it may assume such dimensions that the penis becomes so involved that phymosis and paraphymosis may supervene, to add to the other complications. These, however, are not serious sequelæ, as by proper care, with fomentations or scarifications, and the use of a suspensory bandage, they may be readily overcome.

GANGRENE.

This accident may be looked for from the fourth to the eighth day, manifesting itself not only by the extent which the œdema of the scrotal region assumes, and by its characteristics of coldness, loss of sensibility, and crepitant feeling, but by the fœtid odor proceeding from the wound, and by a change in the character, or the disappearance of the suppuration, which is succeeded by a sanious, bloody and offensive discharge. To this series of symptoms are to be added a marked increase of the general disturbance, manifested by increased thirst, anorexia, fœtid mouth, change of color in the mucous membrane to a livid hue, increase of pulse, with weakening, increased respiration, temperature at first elevated and then diminished, and after five or six days a final termination in the death of the patient.

The progress of this complication is so rapid, and the chances of recovery are so few, that the necessity for prompt treatment becomes at once obvious. All the diseased and mortified parts must be removed at once, and means instantly employed to prevent the absorption of gangrenous matter. Friction with ammoniacal and turpentine liniments must be used over the swelling; the parts must be subjected to the actual cautery at white heat, and disinfecting agents of all kinds must be freely used, as chloride of lime, carbolic acid, and permanganate of potash, while internal treatment must immediately be instituted by the administration of stimulants and antiseptics in the form of ammonia and phenic acid, or its preparations.

ABSCESSES.

When these are likely to result from a too rapid closure of the edges of the scrotal envelope, the premature union may be readily prevented, as we have before stated, by the careful introduction of the finger into the wound while it is still suppurating. But notwithstanding this precaution they will sometimes occur as the result of the infiltration and accumulation of the suppurative matter. A free incision and proper attention to the cavity of the abscess, is all that this accident requires. A careful examination of the parts will, however, reveal another cause for the formation of these abscesses. It is then against these causes that the therapeutic treatment must be directed. We refer now to the complication known as the formation of a

CHAMPIGNON.

This name is applied to an indurated condition of the end of the cord, or in its thickness, of a tumefied character, varying in size and extent, and slow in its growth. It results from an excess of inflammatory action, attributable to the manipulations which become necessary during the performance of the operation. The name "champignon" (or mushroom) is applied to it by the French, on account of the pedunculated appearance which it sometimes assumes, and which causes it to greatly resemble that fungus in its outward figure. It is also known as schirrous or indurated cord. The tumor is sometimes situated on the outside of the envelopes, when it is known as *extra-scrotal*, but more commonly it is found covered by the skin, in which case it is better known as *intra-scrotal*. In this latter condition it may be merely a growth at the end of the cord, becoming, as determined by its location, of an *extra-inguinal* character, or if the diseased process extends as far as the upper inguinal opening, or beyond it, it becomes, and is so denominated, *intra-abdominal*. There is also an *extra-intra-scrotal* growth, when it is partly within and partly external to the scrotum. This tumor will vary greatly in size, being sometimes very small in dimensions, and at others having those of a man's fist. We have ourselves observed it equalling a child's head in size.

The causes from which it originates are obscure, and cannot be very well defined. Still, they may be arranged under the heading of any of the morbific causes which may excite an excess of inflammatory action at the end of the cord. Amongst these may be enumerated all violent tractions upon the cord at the time of the operation; all unnecessary manipulations during the process of cicatrization, such as the too frequent introduction of the finger into the wound with destruction of the granulations already adherent to the cord, and the application of the appliances for its division too low down upon it, leaving that organ hanging too much, and the retraction of the organ being insufficient to retain it in the inguinal sac. Still, as a champignon may be developed in the absence of all these causes, it would seem that their growth may be attributed also to some specific idiosyncrasy in the animal affected, the true nature of which cannot be very accurately or easily understood. It is held, however, by certain German and

Russian authors that exposure to cold exercises a great deal of influence in the development of this affection, and observation has largely established the fact of its greater prevalence during cold seasons.

Symptoms of extra-scrotal champignon. — This is otherwise known in the terminology of some pathologists as *true* or *superficial champignon*. It develops itself at the cut extremity of the cord as a granulating mass, of a red color, varying in size, its growth, nevertheless, allowing the cicatrization of the skin to progress in such a manner that it forms a point of attachment from which the tumor seems to proceed. This form of it is usually of little account, as it may easily be removed before it has attained to troublesome dimensions. When of considerable proportions, however, it may interfere materially with the act of locomotion by causing pain in the cord, upon which it drags more or less. It is not often or necessarily accompanied by constitutional disturbance, excepting in cases of excessive suppuration, which may sooner or later undermine the general health by exhausting the stamina of the patient.

If instead of showing its greatest development on the surface of the scrotum, it occurs beneath it, a greater or less degree of swelling will appear on one or both sides of the inguinal region, the swelling being somewhat hard, possibly the seat of one or more fistulous tracks resulting from abscesses which have at times opened, discharged, and closed; the animal showing a certain amount of stiffness in the action of the hind legs. In this case we shall have to adapt our treatment to the *deep champignon* of Zundel, under one of its three forms of *extra-inguinal*, *intra-inguinal*, and *intra-abdominal*.

Under the first head we shall often discover, upon inquiring into the history of the case, that for a length of time, varying from months, perhaps, to years, the animal had been affected with a swelling which would gather, break, and slowly heal, leaving no mark as an apparent indication of a diseased condition, excepting that a certain degree of lameness would have been observed to be present. Upon exploring the testicular region it would then be observed to be the seat of a tumor, either spherical or pysiform, seldom painful, and more or less adherent to the envelope that covered it. Above this the end may be felt free from diseased process, and this is the champignon in its chronic form. In this condition it is not incompatible with the general health of the animal affected, and forms no hindrance to his usefulness. This condition of extra-inguinal growth will sometimes dissolve away by an abscess-formation, and quite disappear. But if the induration of the spermatic cord extends to the upper portion, or that which is enclosed in the inguinal canal, in such a manner as to interfere with locomotion, the leg corresponding with the diseased side being carried in abduction, with numerous fistulous tracks existing on the surface of the scrotum, the intensity of the symptoms varying with the extent of the diseased process, the condition of the cord will be easily discovered by an examination of the parts, and the presence of an *intra-inguinal champignon* established. If, besides these symptoms, we discover by rectal examination that there is in front of and above the pubes a tumor more or less ovoid, or giving the sensation of a cylindrical mass, of size varying to the touch—which is the diseased indurated cord—the case is judged at once to be one of *intra-abdominal* nature. At times the inflammation may extend to the sub-lumbar region, when the hand introduced into the rectum may discover in that locality an ovoid tumor or abscess which may be of great size. This form of champignon is incomparably the most serious of them all; an intense and presistent react-ive fever is always present, and this at length termi-nates together the life and suffering of the animal. The abscess may sometimes open externally, and in some cases it may accumulate within the thickness of the cord and form large collections; or, again, it may find its way into the abdominal cavity, where it may excite a fatal peritonitis.

This rapid examination of the various forms of deep champignon will enable us easily to realize the difficulty of the progress in the case. While the pedunculated form, exterior to the scrotum, is not, comparatively, a very serious matter, it becomes, on the contrary, a very grave occurrence when it assumes the characters of the intra-abdominal variety, and must in a majority of cases be recognized as an incu-rable disorder.

Treatment.—While champignon is an affection in which surgical interference cannot usually be dispensed with, it is still essential that the surgeon should avoid being over hasty in determining in favor of an operation, and he should give the case a very careful consideration before deciding upon his course. At first emollient applications, appropriate topical treatment, and a few points of cauterization, may be followed by a process of resolution. But in the event of their failure four modes of operation present themselves. These are, in their order, the application of the clamps; the ligature; the linear crushing or ecraseur; and cauterization. When the case is one of the extra-scrotal variety, the application of a ligature around the base of the peduncle, or removal by ecraseur, will be the simplest mode of treatment, unless there should exist a tendency to infiltration of the cord, in which case the manipulations to be followed become the same as those which are adapted to that of the deep or intra-scrotal form.

By the clamps.—When the application of the clamps is resorted to, they may be similar to those used in

FIG. 25.

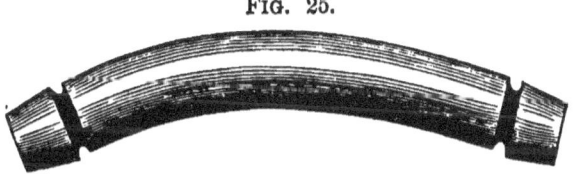

CURVED CLAMPS.

ordinary castration, or may be curved in form (Fig. 25). The animal to be operated on is to be thrown on either side according to which cord is affected, and an incision made through the envelopes as nearly parallel with the median line of the body as the case permits, when the tumor and the cord are carefully dissected and separated from their adhesions. If the tumor is suspended from the end of the cord there will be no difficulty in applying the clamp above it and upon a healthy portion of the cord. But if the diseased process extends within the inguinal canal, there will be need of great caution in dissecting the cord up to the healthy structure. In doing this the safer mode will be for the operator to treat the adhesions with the fingers or the blunt end of the scissors, rather than to employ the sharp edge of the bistoury with the accompanying danger of causing hemorrhage. If, on the contrary, the cord is diseased to an extent that renders it difficult to reach a healthy portion, other modes of operation—as by the ligature—become the wiser and more practicable indication. When the clamps are used it is necessary to leave them in place for several days, and sometimes they are allowed to slough off, while the growth is usually suffered to remain for a few days after the operation.

Ligature.—When this mode of procedure is adopted, the tumor having been dissected and the cord well freed from its adhesions with surrounding parts, and the ligature being applied, the tumor may either be amputated immediately or be left to slough off in its own time. The ligature may be either of twine, silk, or elastic cord. We have ourselves operated by this method in the successful removal of growths of very considerable size. So long as the upper portion of the cord, which retains its healthy structure, can be reached, the application of the ligature is attended with no difficulty, the manipulations required being similar to those which attend the removal of all growths by the process of ligation. But if the diseased process extends so far that the ligature cannot be applied at the proper point, as in the case of intra-inguinal champignon, it will be necessary to have resort to the ligature-carrier recommended by Serres (Fig. 26). In using this instrument the loop

FIG. 26.

LIGATURE-CARRIER.

of the ligature being passed over the tumor around the cord, is carried into the inguinal canal as high up as possible, pressure being made by holding the instrument against the cord, while strong traction is made on the ends of the ligature, which is then secured by a knot upon a small stick placed across the opening of the instrument, with a view to the prevention of slipping. If an increase of pressure is found to be necessary, it can easily be obtained by tightening the ligature from day to day as required. If the size of the cord should be such as to prevent a proper application of a single ligature, it may become necessary to divide it in applying a double, triple, or multiple ligatures, in accordance with the rules for such ligating.

In whatsoever manner the ligature may be applied, even when it is of the elastic kind, the process of sloughing of the tissues is always a slow one. It is for this reason that we agree with Prof. Bouley in considering the treatment of champignon by the linear crushing very much to be preferred.

Ecraseur.—The steps of the operation with this instrument are similar to those required in the other methods already considered. The champignon is isolated from its surrounding parts, the chain is placed on the cord above the base of the tumor, and the amputation is completed by a slow pressure upon the cord, which, crushing it by degrees, permits its immediate removal. It must be done slowly, occupying from ten to twenty minutes for the complete separation of the champignon, according to the size of the tumor. The operation being finished, the parts are left in the condition of a simple wound, where no cause exists to interfere with its rapid cicatrization.

Cauterization.—This is a mode of treatment which we have never had occasion to submit to trial, having always given the preference to the process we have just referred to. It is recommended, however, by European authorities. Some of these advocate the "melting" process, or the introduction of sharp points deeply into the thickest parts of the enlargement, while others advise a removal of a portion of the growth and deep cauterization afterwards. If cauterization can be advantageously employed, the best method, in our judgment, would be the process of amputation with Paccalin, or with the galvanic cautery. We may here, while referring to the application of electricity in this connection, appropriately refer to our own experience of a number of years ago, in treating an animal suffering with intra-scrotal champignon, by electrolysis, and succeeding after two applications, in obtaining the complete removal of the tumor. This method, however, consumes too much time to justify its employment in general practice.

FISTULA OF THE SCROTUM.

Being already aware of several causes of this complication of the operation of castration, we may readily appreciate the treatment they require. It must be remembered that in a majority of cases, the cause of this lesion is the presence of a foreign body in the wound, and that until it is removed, it is in vain to look for a cure. Prof. Bouley has reported a case in which the fistula was due to the presence of a pair of clamps over which the skin had almost entirely cicatrized.

INGUINAL HERNIA; HERNIA OF CASTRATION.

By this is understood the protrusion of some portion of the contents of the abdominal cavity through the inguinal ring, either a portion of the omentum or of some part of the small intestines, creating either an epiplocele or an enterocele. This complication may take place either during the operation, or shortly afterwards, or at the period of the removal of the clamps. It proceeds from the violent struggling of the animal during the operation; to the colics which are so apt to supervene; to his position when placed in a stall of which the floor is too much inclined; or it may result from some of the various modes of castration, as, for example, the uncovered operation.

At times the two forms of hernia may present themselves together, constituting a case of entero-epiplocele. When the epiploan alone protrudes, it need not give rise to any unnecessary anxiety, as it may easily be either reduced and returned to its place, or ligated with the clamps, or torn apart. If, on the contrary, it is a portion of the small intestines which becomes involved, the first indication is to restore it to its place by the proper taxis without delay, which may be readily done, the animal being yet down and placed under an anesthetic, by the rectal taxis combined with the necessary inguinal manipulations. When this has been accomplished the intestine is kept in place by the application of a clamp over the cord, upon which the fibrous coat of the cremaster has been carefully drawn.

PERITONITIS.

This complication, considered as one of the most frequent following castration, is also, beyond doubt, one of the most serious. It is generally the result of exposure to cold, especially when its occurrence accompanies the suppurative fever. But it also develops itself in animals which have received the best hygienic care, its appearance being attributed to an excessive dragging of the cord, or to the extension of the local inflammation by continuity of tissues. It manifests itself generally between the second and third day following the operation, except when it becomes symptomatic, as of gangrene of the cord, when we have seen it making its appearance towards the tenth day.

The symptoms of this traumatic peritonitis differ somewhat from those of the acute inflammatory type. According to Gourdon, "the animal is dull and refuses all food—the suppuration of the wound of the scrotum has ceased, the bags and surrounding parts become the seat of a warm, hard and painful swelling. The animal stands with his four legs brought close together, the back is stiff and arched, the flanks are cordy, the abdomen painful, the pulse hard, small and increased. As the disease progresses, the symptoms are more marked, the enlargement of the envelopes increases and is more diffuse, it extends down to the abdomen, and even under the chest, passes along the thighs, is less warm, less hard, less painful, and pits under pressure. There are slight colics, the pulse gets smaller, intermittent, the respiration is increased, and the animal dies towards the fifth or sixth day."

The treatment to be recommended varies according to different authors. While some prescribe depletive and sedative treatment, laxatives and diuretics, many prefer tonics and stimulants. The Germans claim great results from the use of tincture of arnica (in small doses) administered internally. The external treatment consists in sinapisms, warm fomentations, poultices, or fumigations under the abdomen.

TETANUS.

As with most cases of traumatic tetanus, this complication is generally fatal, and it is, without doubt, the most dangerous of all and marked by the greatest mortality. It is generally admitted that exposure to cold and dampness is one of the most prolific causes, especially in animals which, having but recently recovered, are too soon put to work. The various modes of operation have also been considered to have some influence upon its development, though there is probably no ground upon which this theory can find a support. Whether the nature of the soil of a district, or its atmospheric condition, may have any connection with it, is also a question. We know that in some portions of Long Island, cases of tetanus are commonly met with, at some seasons of the year, after surgical operations of every kind. It may appear within a few days following the castration, or it may defer its visitation for a period of twenty days, or longer.

The treatment adopted for the tetanus of castration is that which is applied to all cases of that traumatic affection.

AMAUROSIS.

This disease may also be included among those classed as the sequelæ of castration, having been known to follow cases where hemorrhage of the small testicular artery had occurred. Tonic treatment internally and local stimulating applications may sometimes relieve this complication, but it will generally be admitted to be incurable.

COMPARATIVE VIEW OF THE VARIOUS MODES OF CASTRATION.

The process by *simple excision*, by reason of the hemorrhage which necessarily accompanies it, though not inevitably dangerous, must be excluded from the domain of general practice.

That of *scraping the cord* has not, so far as our knowledge extends, been sufficiently tested, either in European or American practice, to justify its recommendation.

The process of *torsion below the epididymis* is too much subject to the development of champignon, as well as that of *free torsion* with the hands, to be admitted by judicious operators, while the *limited torsion* is a method which has taken rank amongst safe operators, notwithstanding the enormous swelling of the parts by which it is commonly accompanied, and the necessity it involves of the introduction of the fingers into the wound to prevent its premature closing.

The method by the *ecraseur*, though occupying a longer time in its completion than some others, has secured very favorable results, especially in the hands of American operators.

The operation by *cauterization* is highly recommended by English veterinarians. We believe, contrary to the statements of French authors, that it is not widely in use on this Continent. The objections urged against it are that the hemostatic effect upon the cord is less reliable than in the method by the clamps or the ligature; that there is more or less danger of cauterizing the surrounding parts by the effect of the radiant heat from the cautery; and that the swelling which follows the operation is always excessively great.

Castration by the *clamps* is the best known and most extensively practised. It is easy and quick in its performance; performs the most certain hemostasis upon the artery, and notwithstanding some slight objections, merits a preference over all others. The principal objection alleged against it is that it is attended with great pain to the suffering patient when the pressure of the instrument upon the soft tissues is first felt. This is a doubtful question, and if this excessive amount of pain really exists, it certainly cannot be of long continuance, merely on account of the effect produced by the clamps themselves.

Of the various methods by *ligature*, that of the ligation of the cord with its envelopes is applicable to small animals only. That upon the cord alone is liable to be followed by hemorrhage, or by the excessive retraction of the cord into the abdominal cavity, drawing the ligature with it. That of the efferent canal, and of the cord by the subcutaneous mode are not admitted in general practice, while that of the artery alone has not been extensively performed on large animals, so far as we are informed, except by certain Massachusetts veterinarians.

The castration by *double subcutaneous twisting*, when extensively applied to solipeds, will probably prove to be the safest mode of all, and least likely to be followed by complications. We are not informed as to the extent to which it has been practised in this country, even amongst ruminants.

CHAPTER VI.

CASTRATION OF FEMALES—HISTORY—INDICATIONS—EFFECTS UPON THE ORGANISM AND SPECIAL FUNCTIONS—ADVANTAGES IN COWS—CONDITIONS FAVORABLE TO THE OPERATION—ANATOMY—MODUS OPERANDI—BY THE FLANKS—CHARLIER'S PROCESS—INSTRUMENTS—VARIOUS STEPS—DIVISION OF THE VAGINA—SEIZING THE OVARY—TWISTING IT OFF—COMPLICATIONS—HEMORRHAGE—PERITONITIS—ABSCESS OF THE PELVIC CAVITY—CONSTIPATION—SUBCUTANEOUS EMPHYSEMA—CASTRATION OF THE SMALL ANIMALS—OF SWINE—OF SLUTS—OF FOWLS.

As I have stated before, the revival of the operation of castration upon large females is due to a Louisiana farmer, Thomas Winn, who, in the year 1831, castrated several of his cows. Without entering upon the history which includes a record of the failures and successes attendant upon the introduction of the operation, it may suffice to say that until the improvements made by Charlier in the manipulations involved in the operation, it encountered considerable opposition, and it is within a comparatively recent period that it has become established in the domains of veterinary surgery.

The indications by which this operation commend itself to agriculturists, and others who find profit or pleasure in the use or ownership of these domestic animals, are several. Among them are the influence which it exercises upon the secretion of milk in cows, and upon the power of accumulating fat, and its effects upon the character and temper of all the large females, in which relation it obviously acts as a therapeutic agent, in overcoming certain peculiar conditions by which they are distinguished. In respect to the effect of the operation of spaying the cow upon the milk secretion, it is a fact well established that it not only increases the amount and duration of the flow, but also improves the quality of that valuable fluid, the spayed cow not only continuing the production from eighteen to twenty-four months, but giving a product far richer in the elements of nutrition. This is shown by the enhanced proportions of the cream, the caseine and the sugar, which determine its richness and value, both economically and commercially, after alteration.

But even this argument in favor of spaying the cow is rendered more weighty by the fact that besides its influence on the milky secretion, there is also that which is furnished by the consideration of its effect in augmenting the deposit of fat throughout the frame, for it is through this tendency that the flesh of the animal becomes so greatly improved in its nutritive quality as compared with that of the same species when in the entire condition, becoming so noticably more tender, juicy and palatable, retaining more of the oily element, digesting more easily, and so, of course, acquiring a pecuniary value in the market not before possessed. These remarks apply to the dry equally with the milch cow, and leaving out the reference to the milk secretion, to the ox as well.

With respect to the effect of the operation upon the character and disposition of the cow, these are easily illustrated in the movements of the nymphomaniac animal, which may be said to be constantly in a state of hysterical excitement. They seem to be in continual conditions of heat, running after and mounting other animals with which they may be in company, while never producing and giving no milk. They are always in a lean condition, and must remain a pecuniary loss to the dairyman. This manifestation of nymphomania is also met with in the mare, which, continually exhibiting signs of heat, becomes more or less dangerous on that account. In these cases the advantage of the operation of spaying cannot be overlooked. We have personal knowledge of several cases of this character, in which worthless and troublesome mares have been transformed into docile and valuable animals.

CONDITIONS FAVORABLE TO THE OPERATION.

Charlier expresses the opinion that the best time for the performance of the operation upon cows is from the sixth to the eighth year, or after they have had their second or third calf. If performed at an earlier period the great objection originally urged against castration, that its performance would tend to the diminution of the stock in numbers or "population," might find more or less confirmation. But by an observance of this condition all danger of the annihilation of stock would be obviated. The cow to be operated on ought to be in fair condition, not in heat or pregnant, and the time selected should be from forty to sixty days after calving.

ANATOMY.

The *vagina* is situated within the pelvic cavity, between the rectum and the bladder. Its internal face presents numerous longitudinal folds, the purpose of which is to permit the free dilatation of the parts. At the bottom of the passage is situated the *neck of the uterus*, giving to the finger the sensation of a projection, hard towards the cavity of the vagina, and in the centre of which is felt a closed opening, from which radiate the folds of the mucous membrane. The *uterus* (Figs. 27 and 28), continued forward to the neck, is situated in the abdominal cavity, occupying the sub-lumbar region, with its posterior extremity resting at the end of the pelvic cavity. It is somewhat pyriform in shape, and larger at its base, where it divides into two lateral halves, continued by the *horns*. The concave curvature of these horns look downward in the cow, but face upwards in the mare. In both they give attachment to the *broad ligaments*. These are folds of the peritoneum, more developed forward than behind, rising from the sub-lumbar region, and descending towards the uterus, to fix themselves upon the sides of the superior face of the body of this organ, and, as before stated, upon the curvature of the horns. Their anterior border is free, and gives support to the oviducts and the ovaries. Between the serous layers are found the utero-ovarian artery and veins, largely developed. The *ovaries* are situated on the internal face of the broad ligament, forming a small ovoid mass, which receives a special serous lamella, a sort of ligament, having between its layers a few grayish muscular fibres, which may be strong enough to offer serious resistance when the extirpation of the organ is attempted.

MODUS OPERANDI.

There are two modes of operation. The original method was that of removal through the flanks, which, however, has fallen into disuse since the introduction of the process of Charlier, of removal through the vagina. This process is altogether to be preferred, as being safer, more consistent with scientific surgery, and in a word is the only one which it is proper to perform, so long as the capacity of the vagina permits the necessary manipulations to be performed.

METHOD BY THE FLANKS.

Four steps are necessary to be followed in this

Fig. 27.

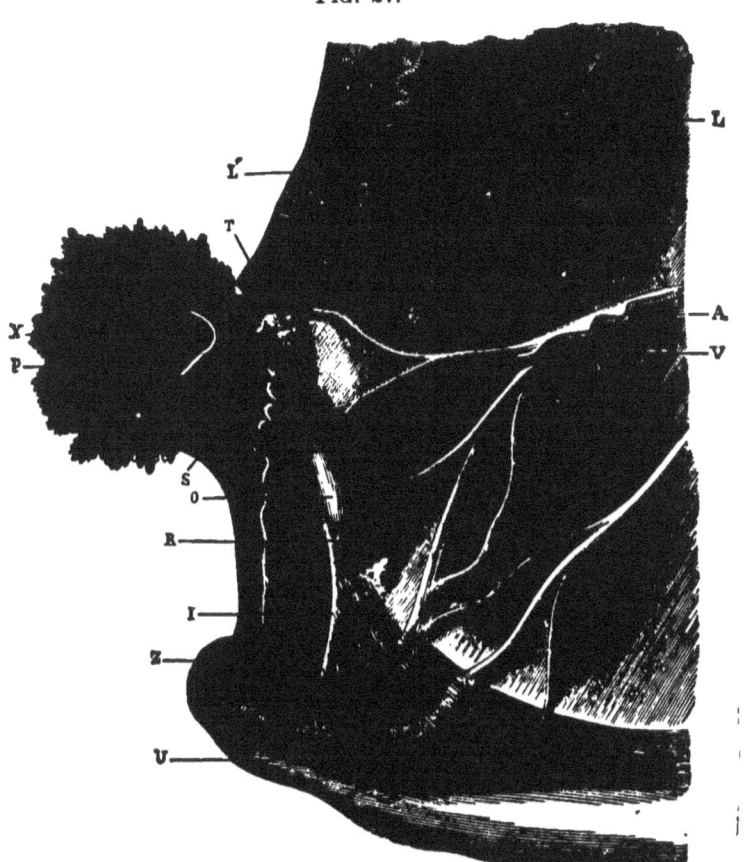

RIGHT OVARY OF THE COW WITH ITS ATTACHMENTS.

U.—Right horn of the uterus. L.—Broad ligament. L'—Its anterior border. O.—Ovary. R.—Peritoneal fold where it is suspended. S.—Superior ovarian ligament. T.—Inferior ovarian ligament. A.—Ovarian artery. V.—Ovarian vein. I.—Oviduct. P.—Its pavilion. X.—Its superior or fimbriated opening. Z.—Its inferior opening.

FIG. 28.

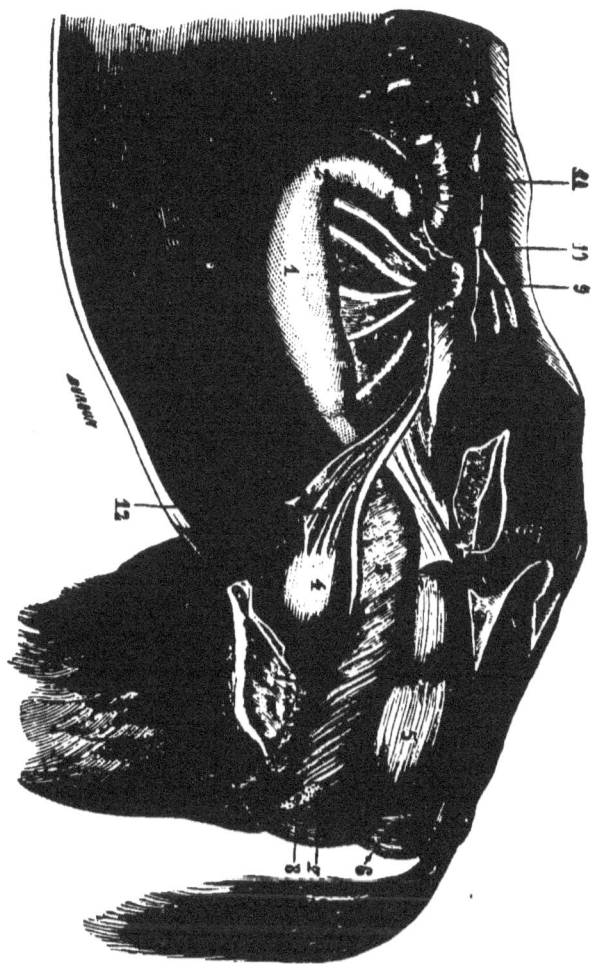

LONGITUDINAL SECTION OF THE PELVIS OF THE MARE SHOWING THE POSITION AND CONNECTION OF THE GENITAL ORGANS.

1—Uterus. 2—Horn of the uterus. 3—Vagina. 4—Bladder. 5—Rectum. 7 8—Vulva. 9—Ovary. 10—Oviduct. 11—Kidney. 12—Broad ligament.

method. The first is securing the animal. The cow is usually kept on her feet, pressed firmly against a wall, the legs secured with hobbles, and her head controlled, as much as possible, by a strong assistant. The second step is the incision of the flank. This is made on the left side, with a sharp, convex bistoury, in the middle of the superior portion of that region, dividing the skin and muscles vertically, care being taken that the incision is not carried too low down, in order to avoid the division of the circumflex artery, which passes along in that vicinity. An opening is then made in the peritoneum, either with the knife or with the fingers, sufficiently large to permit the introduction of the fingers. In the third step of the operation, which comprehends the removal of the ovary, the surgeon introduces his hand into the abdomen, and turning it towards the pelvis, feels for the horns of the uterus. Upon finding these the ovaries are easily discovered. He carefully draws them outwards, and their removal is effected either with the ecraseur or the forceps of Charlier. The operation is concluded by the application of a quill suture.

CHARLIER'S METHOD—INSTRUMENTS.

For this operation special instruments are required. These consist of, first, a vaginal dilator (Fig. 29), or speculum, of peculiar and somewhat complicated construction, to be modified subse-

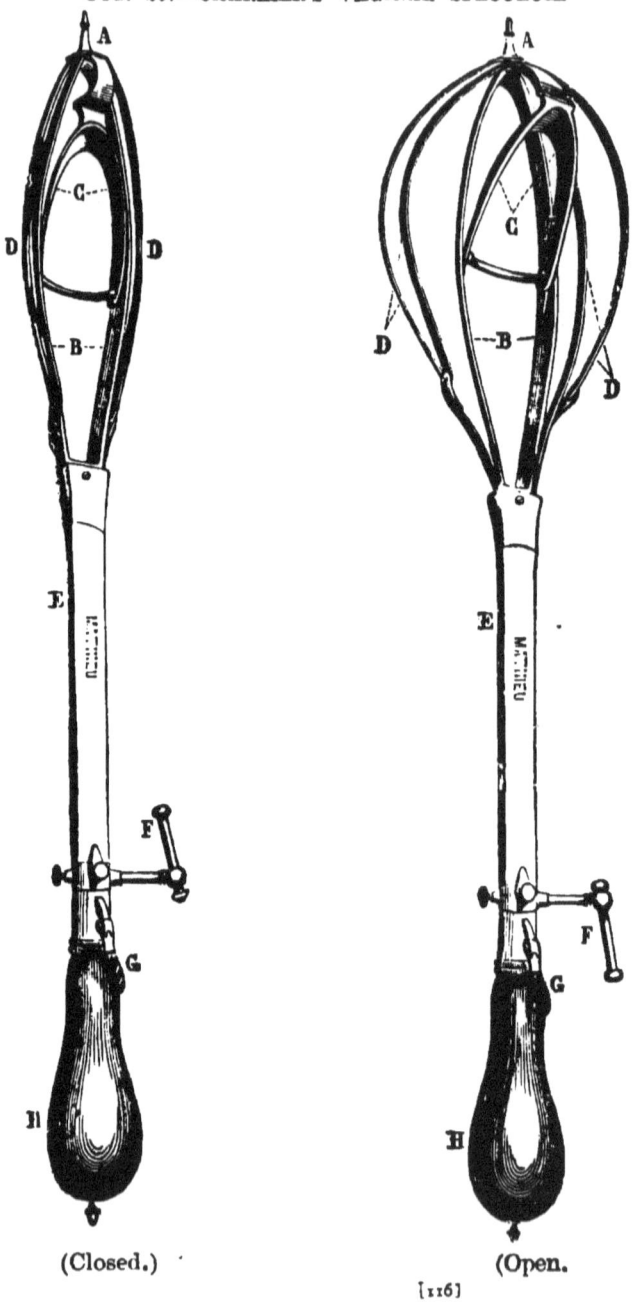

FIG. 29.—CHARLIER'S VAGINAL SPECULUM.

(Closed.)　(Open.)

ANIMAL CASTRATION. 117

Fig. 30.

MODIFIED VAGINAL SPECULUM.

quently by another (Fig 30), of superior form and easier of application, and now in general use; second, a bistoury caché (Fig. 31), sliding on its handles,

Fig. 31.

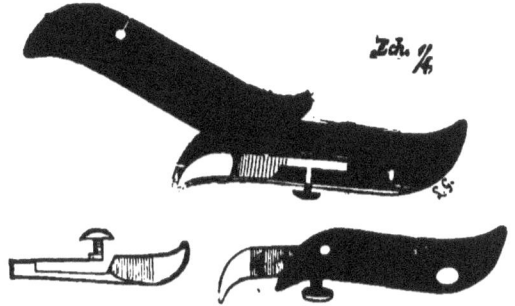

BISTOURY CACH

which is a true embryotomy knife, modified by Colin (Fig. 32); third, a pair of

FIG. 32.

COLIN'S BISTOURY CACHÉ.

long, sharp scissors (Fig. 33), with guarded blades, curved on its flat surface; fourth, a torsion forceps (Fig. 34), closed by a peculiar thread arrangement, moved by the handle; and fifth, a steel thimble (Fig. 35), which has been modified by the instrument shown in Fig. 36, and which

Fig. 33.

Fig. 34.
FORCEPS FOR CASTRATION OF COWS.

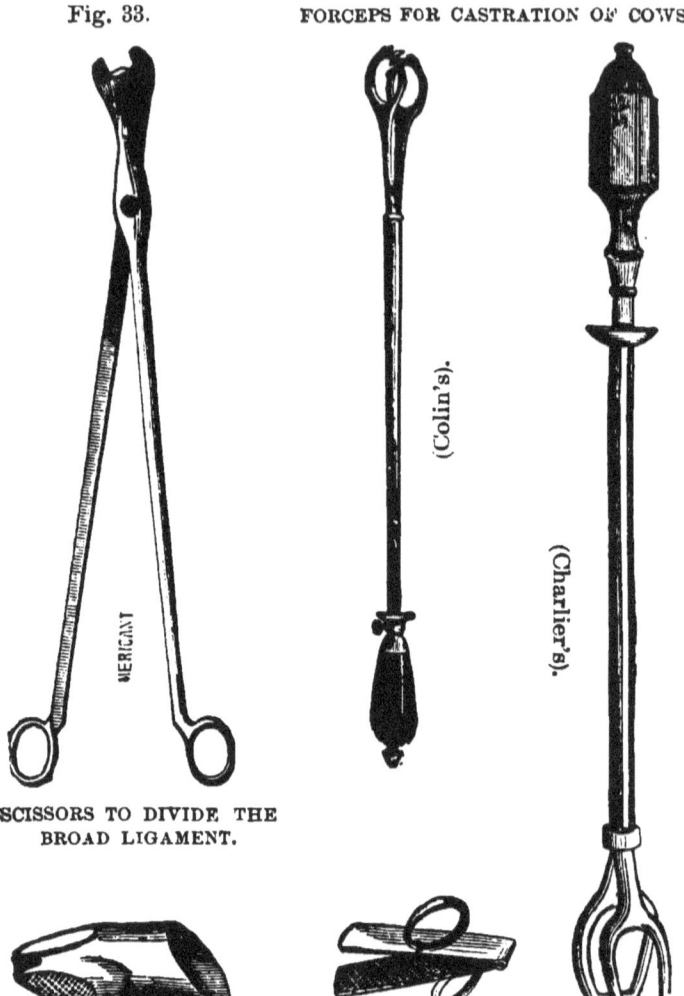

SCISSORS TO DIVIDE THE
BROAD LIGAMENT.

FIG. 35.—THIMBLE
FOR CASTRATION.

FIG. 36 —COLIN'S NIPPERS,
to take the place of the thimble.

is used in applying the limited torsion on the broad ligaments.

Preparation of the animal.—No general preparation is required, except one, which may be regarded as of local effect, but is not to be overlooked. This consists in the evacuation of the bowels by means of a rectal injection, in order that the arms of the surgeon may not become unnecessarily soiled during the operation. The animal is secured on her feet by being placed in a narrow stall to prevent her from moving from side to side, the floor of the stall having an inclination forwards, in order to prevent the pressure by gravitation of the intestinal mass towards the posterior parts of the abdomen.

The operation is completed in two steps, of which the first is the incision of the vagina, and the second the extirpation of the ovaries.

THE INCISION IN THE VAGINA.

This is made in the following manner. The operator introduces the speculum with his right hand, through the vulva, into the vaginal cavity, and carefully passing in his left hand, well oiled, directs and introduces the little prolongation A of the speculum into the centre of the neck of the uterus, gently pressing upon it in order to keep it in place. In using the original dilator, the opening of the branches must be so regulated as to put the walls of the vagina upon the stretch. Or, if he uses the modified speculum, he pushes the instrument downwards and forwards, and by this motion distends the upper wall of the cavity, keeping the instrument in that position by a hold of the left hand, which has been withdrawn from the vagina. He then arms himself with the bistoury caché, which he holds closed in his full hand and introduces with the right hand into the vagina. Carefully feeling the condition of the upper wall of this cavity, and assuring himself of its being well stretched, he rests his hand, still holding the bistoury, upon the opening or "window" at the end of the speculum (Figs. 37 and 38), and by firmly pushing the blade (the sharp edge being turned backwards) out of its handle, pierces with it the vaginal walls, about two inches above the neck of the uterus, and with a motion from below upwards and from before backwards, makes an incision on the median line, from three to three and a half inches in length. The introduction of the instrument must be made in such a manner that it will pass at once through the walls of the vagina proper, as well as through the peritoneal cap which it presents at its anterior portion.

The incision being completed, the speculum is carefully withdrawn; and if a slight hemorrhage should occur, the blood should be removed before the surgeon proceeds to the second step of the operation.

REMOVAL OF THE OVARIES.

Then, again introducing his hand into the vagina, and passing his finger through the opening made by the incision, he feels for the ovaries, which he finds floating at the extremity of their ligaments, towards the entrance of the pelvis, below, on each

FIG. 37.

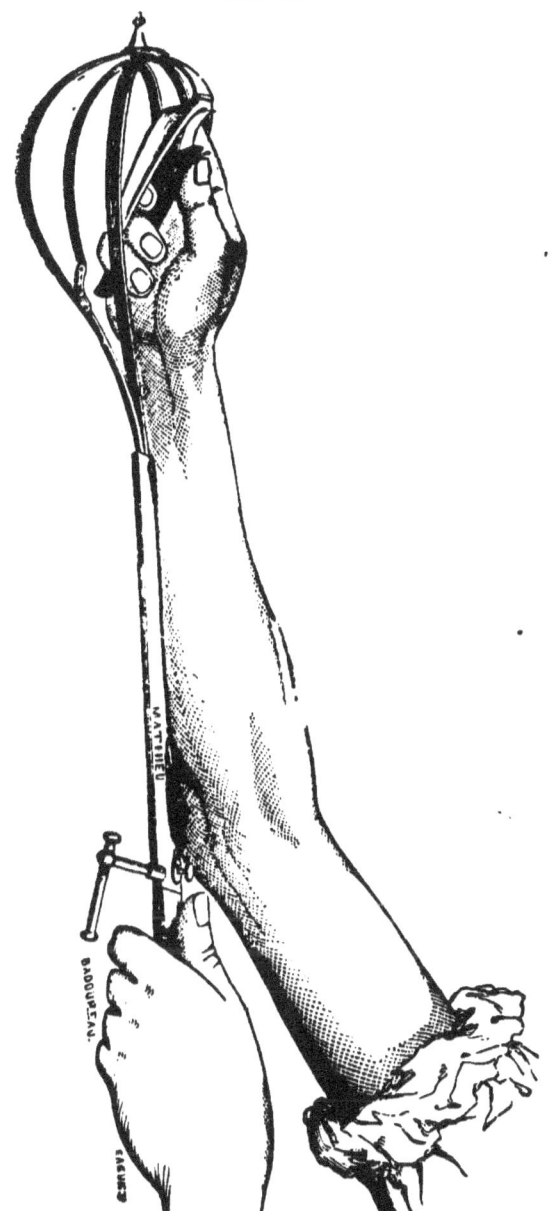

CASTRATION OF COWS. (Charlier's process).
Incision of the vagina.

FIG. 38.

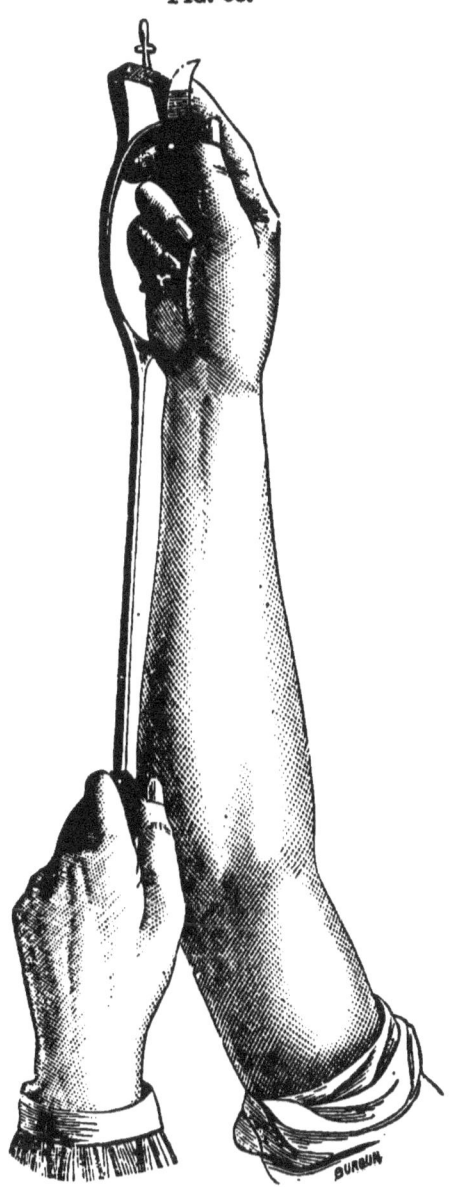

CASTRATION OF COWS. (Charlier's process).
1st step. Incision of the vagina.

side, and at a short distance from the incision, between the base of the uterine horns, near and inside of the free border of the ligaments, and a little above the anterior border of the pubis. Then, grasping the ovarian body, he draws it carefully into the vagina, through the incision, and introduces the long scissors, with the guarded blade of which he divides the thick border of the broad ligament (Fig. 39), replacing the ovaries into the abdomen without releasing his hold. The torsion forceps is then passed into the vagina and through the incision, and is made to take hold with its open jaws of the broad ligament (Fig. 40), above the ovary, and is firmly closed by the movement of the spiral crank of the handle (D). Both hands being now outside of the vaginal cavity, and the forceps being secured on the broad ligament, held by the right hand, the left thumb is protected by the thimble (E), and the hand once more inserted, to grasp the broad ligament above the point where the jaws of the torsion forceps are placed. In this position the torsion is made with the forceps, the twisting of the ligament being limited by the firm pressure made by the thimble on the thumb with the index finger, or by a pair of crushing pincers. After several turns of the instrument, the ovary is separated from its attachment, and may be brought out of the cavity, still held securely between the oval jaws of the torsion forceps. The same method is applicable to the ovaries of both sides. These rules are subject to more or less modification by indications which may occasionally

FIG. 89.

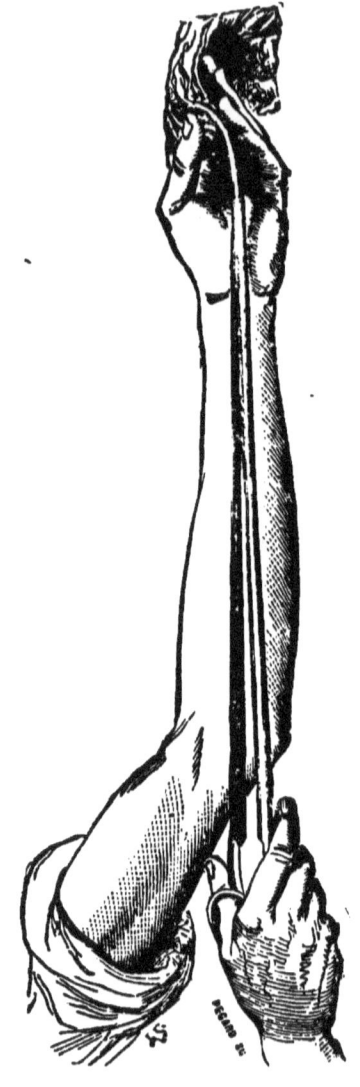

SCISSORS DIVIDING THE BROAD LIGAMENTS.
(126)

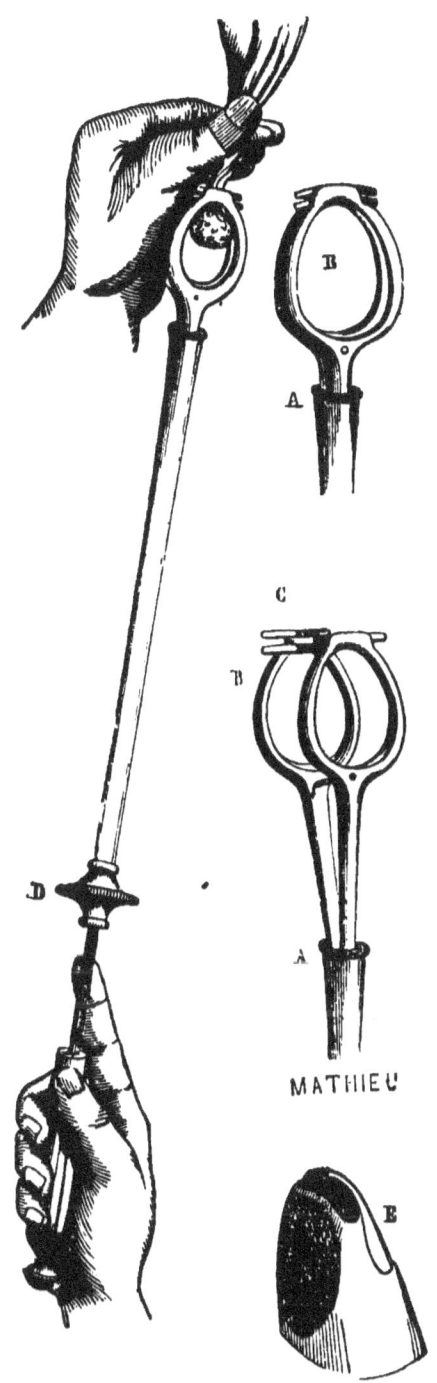

Fig. 40.—TORSION OF THE OVARY.
A B—Forceps closed.   B C—Forceps open.   E—Thimble.

present themselves, arising from the age of the animal or the structure or other conditions of the ovaries.

This method of castration has been modified in many ways, both as to the forms of the instruments used, and the mode of using them, a majority of operators, at the present time even, discarding the dilator, and making the incision simply by stretching the walls of the vagina and pushing against the neck of the uterus with the hand. The manner in which the removal of the ovaries is effected has also been subjected to many changes. For more than twenty years, during which we have been performing the operation, we have been accustomed to use the ecraseur in the last step, and with excellent and satisfactory results.

The subsequent attention required by the spayed cow is usually a very simple matter, and involves little beyond careful dieting, the patient recovering from the general effects upon the system usually in a few days.

COMPLICATIONS.

It may properly be said that there are no complications likely to follow the process in the castration of cows, which may be denominated serious. In the statistics which record the mortality attending it, the fatal cases are represented at the very trifling rate of two per cent. A light colic may sometimes follow it, but it usually subsides without medical treatment. Still, however, spaying may at times be accompanied by accidents of a serious character, though these have considerably diminished in frequency since the introduction of the method of Charlier. One of these is

HEMORRHAGE,

which may occur when the torsion or the crushing of the artery has not been sufficiently complete. But though it is likely to give rise to peritonitis, it is not necessarily a fatal complication. We have ourselves known of cases of its occurrence in mares which had survived it a number of days, and when destroyed exhibited none of the lesions of that affection.

PERITONITIS.

We have several times met with this sequel to the operation, especially in mares. But in these cases, as revealed by *post mortem* investigation, the disease seemed generally to have remained localized. Less common than prior to the practice of castration per vagina, it still is followed by fatal consequences when the entire peritoneum becomes diseased. Its appearance usually occurs from the third to the sixth day. There is suspension of the milky secretion, general dullness, chills, anorexia, suspension of rumination, rapid, small and thready pulse, sometimes painful respiration, rapid loss of flesh, and speedily—death. The indications of treatment are similar to those which are applicable to peritonitis in the solipeds, but the prognosis is always serious.

ABSCESS IN THE PELVIC CAVITY.

This is a complication we have quite often encountered. Besides the general symptoms, there are those of a local character, which are detected by rectal examination, by which discovery is made of the presence of a tumor on one side or the other of the vagina, varying in size, fluctuating, and easily identified. This abscess may be opened in the cavity of the vagina, and should be attended to as soon as discovery is made of the fluctuating character of the growth, without waiting for the process of natural resolution.

SUBCUTANEOUS EMPHYSEMA.

Emphysema of the subcutaneous connective tissue is said to be a common sequel to the flank operation. Its appearance need not excite any special uneasiness, as its termination is usually by spontaneous disappearance. It is an accident we have never encountered in our practice. .

CONSTIPATION.

This complication, which is often met with in mares, is to be carefully looked for, and must be relieved by laxative food and rectal injections. It is due to the pain which accompanies defecation while the wound of the vagina is healing, and which the animal tries to avoid by keeping the rectum full.

CASTRATION OF THE SMALL ANIMALS.

SMALL RUMINANTS.

For these subjects, two modes of operation are to be principally recommended. The first is the double subcutaneous torsion; the other the ligature *en masse* of the cord and its envelopes. Having already considered these operations, a passing reference will suffice here.

FOR SWINE.

In *males*, the varying modes employed are the ligature, limited torsion, and the clamps.

In *females*, it must be remembered that the horns of the uterus are very long and flexuous (Fig. 41), and that the very small ovaries are situated on the inside of the broad ligaments, which are very large, and allow the horns to float freely amongst the circumvolutions of the intestines. The animal must be prepared by being secured upon the right side in order to expose the left flank. The incision is made with a knife of peculiar form (Fig. 42), the

FIG. 42.

BISTOURY FOR THE CASTRATION OF SOW.
(Division of the flank).

coarse bristles having been previously closely clipped off. Care should be taken to carry the left leg in extension backwards, in such a manner that the edges of the various tissues divided shall not meet each other when the operation is completed. The incision may be made either vertically, horizontally, or obliquely. When vertical it should be immedi-

FIG. 41.

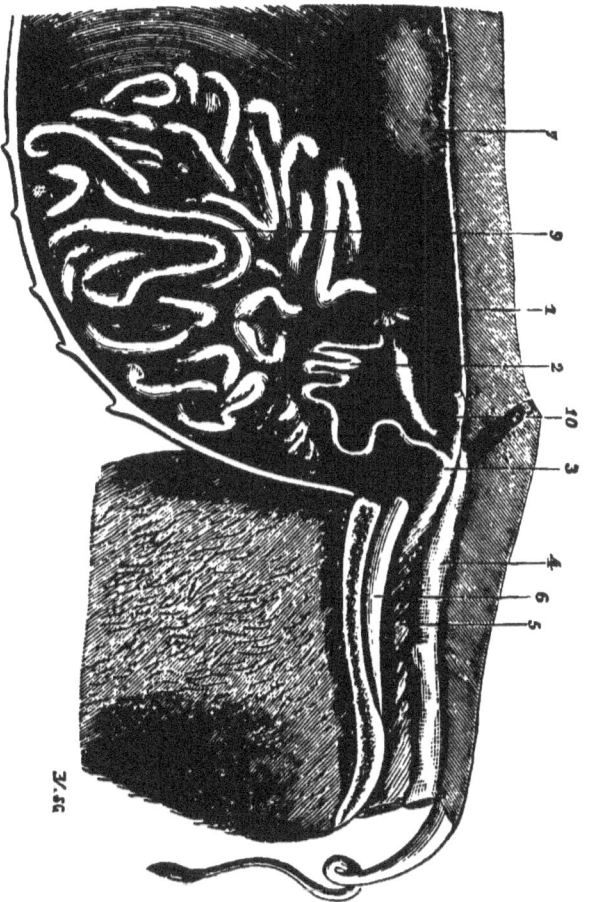

GENITAL ORGANS OF A YOUNG SOW.

(Median and antero-posterior section). 1—Ovary. 2—Horn of the uterus. 3—Body of the uterus. 4—Rectum. 5—Vagina. 6—Bladder. 7—Kidney. 9—Intestines. 10—Stump of one of the horns of the uterus.

ately below the lumbar vertebra next to the last rib; if horizontal it should be parallel with the vertebral column. The vertical incision should be preferred, because it brings the ovaries within easy reach of the fingers. It should be from two to three inches in length, and should be made by a single stroke of the knife, and without dividing the peritoneum, which should, afterwards, be either torn with the finger or carefully cut while raised with the forceps. To find the ovaries the operator introduces the index finger of the right hand between the vertebral column and the intestines, and explores the lumbar region. Upon finding the ovarian sac, he presses it against the abdominal wall and causes it to slide by pushing towards the opening through which it is extruded and grasped. While it is held there the left horn is carefully drawn out after it, until arriving at the bifurcation of the horns at the uterus, the right horn also is brought out and the ovary on that side secured. Both glands being now outside, they are torn or scraped off from their attachment, and the horns are returned to the abdomen.

While this process is readily applicable to young sows, and requires a certain amount of practice to be performed expertly and with success, it is slightly modified when applied to older animals. In that case the two horns must not be exposed outside together, but each must be returned when the removal of the ovary connected with it has been effected.

The simple tearing of the ovaries is not always sufficient, and may be sometimes followed by serious hemorrhage. The scraping and the torsion are safer, and in some instances the ligature has been applied. The incision is closed with the interrupted or, which is preferable, the continued suture. No special after-treatment is required beyond low diet for a few days, with a little extra attention to cleanliness.

The operation may at times be rendered difficult by exceptional and accidental conditions, as, for example, the shortness of the fingers of the operator. This difficulty, however, can be overcome by placing a bundle of straw or other substance under the right flank, which, by raising the body displaces the intestines upwards and crowds the ovary towards the left flank.

It may also happen, as sometimes with old sows, that the ovary has become the seat of large cysts, or that its size is increased in consequence of pathological changes in its structure. In the first case, the cyst may be punctured and emptied with a trochar before attempting the obliteration of the organ. In the second, the opening into the abdomen must be enlarged sufficiently to permit the exit of the extra bulk.

If through inadvertence the operation has been begun while the animal is in a state of pregnancy, the proceeding must be discontinued, the patient kept quiet and the matter indefinitely postponed.

DOGS.

The *male* is altered by either the process of ex-

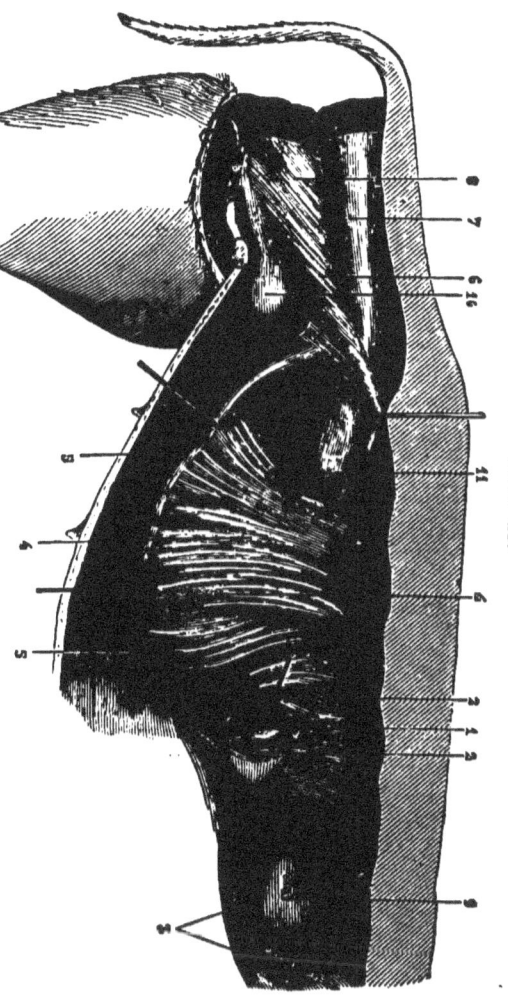

Fig. 43.

GENITAL ORGANS OF THE BITCH.

1—Ovary. 2—Fold of the broad ligament, displaced to expose the ovary. 3—Internal fold of the same. 4—Broad ligaments. 5—Horn of the uterus. 6—Its body. 7—Rectum. 8—Vagina. 9—Kidney. 10—Bladder. 11—Descending colon.

(137)

cision, torsion, or ligature, according to the age and size of the animal.

In the *female*, when, as is sometimes supposed, it becomes a preventive measure in respect to hydrophobia—though if it be so, it can only be from the fact that a castrated bitch will usually remain at home while others are running abroad in heat, and thus being more exposed to contagion—the operation is only justifiable in the case of house dogs, domestic pets, in order to obviate the annoyance caused to their owners by their demonstrations while in heat. In the bitch the broad ligaments are very long (Fig. 43), extending as far as the hypochondriac region, where they divide into an external layer, which reaches to the last rib, while the other extends to the sub-lumbar region behind the diaphragm. The broad ligaments diminish in height as they run forward in such a way that the anterior border of the external layer where the ovary is found, shorter in its median part, gives a certain amount of fixity to the anterior extremity of the horns which it keeps elevated in each hypochondriac region; on that account both horns cannot be at one time brought through the incision, and it becomes generally necessary in the bitch to operate on each side.

The manipulations are similar to those followed in spaying sows, with the exception that the incision is made lower, more forward and nearer to the last rib.

CASTRATION OF FOWLS.

The effect of this operation upon the quality of

the flesh and the power of accumulating fat, in the domestic fowl is a fact too familiar to those who have learned to appreciate the exquisite juicy quality of the meat of the capon to need any comment at our hands. The operation upon these animals is one of considerable difficulty and requires skill and experience to perform with nicety and success.

In birds the testicles are situated in the abdominal cavity, immediately behind the lungs, under the vertebral column and in front of the kidneys (Fig. 44). They correspond exactly to the articulation of the last three ribs with the spinal column, where they lie close together and in contact with the blood vessels which separate them from the kidneys. They are held in position by the peritoneum spread above them, and by minute blood vessels, branches of the aorta or of the vena cava.

In the operation the fowl is placed on his side, the tail being towards the operator, with the hind leg carried backwards, in order to expose the flank of the side selected for the incision. The first step of the operation consists in plucking the feathers from a sufficient extent of surface, and making an incision a little behind the lateral internal processes of the sternum, from within outwards, and from before backwards, and slightly oblique, through the skin and the thin muscles of the abdomen, and when reaching the peritoneum carefully opening it with a puncture, having it raised with a pair of forceps. The second step, or that which involves the extirpation of the gland, is performed by the introduction

FIG. 44.

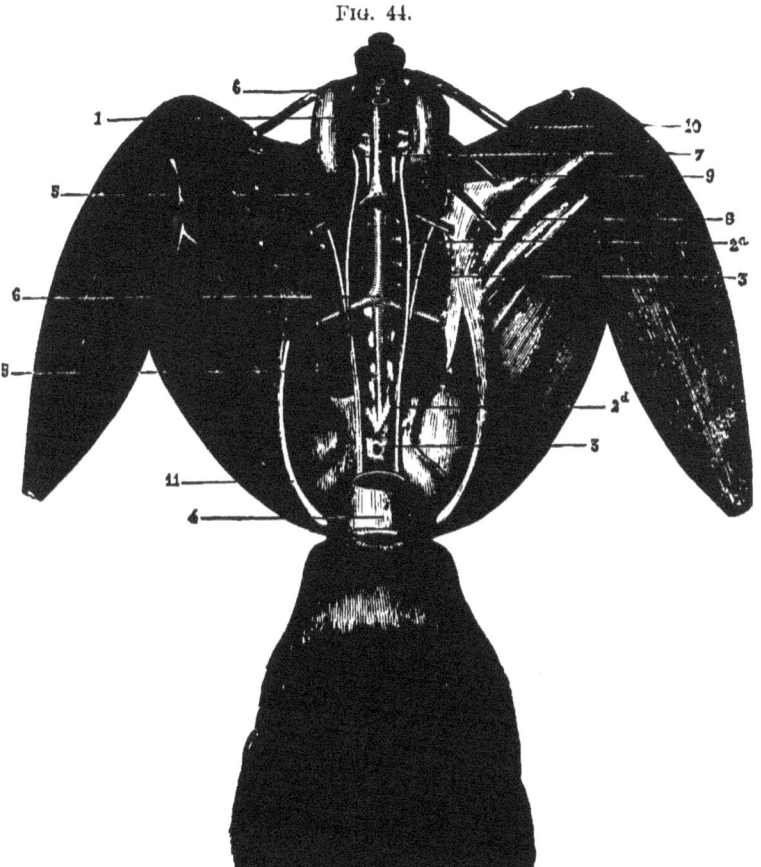

GENITAL ORGAN OF THE ROOSTER.

1.—Testicle. 2a 2d.—Deferent canals. 3.—Ureters. 4.—Cloacum. 5.—Posterior aorta. 7.—Posterior vena cava. 8, 9, 10.—Three last ribs. 11. - Pelvic bones.

of the index finger of the right hand into the abdomen, passing it above the intestinal mass and turning towards the dorsal region near the articulation of the last two ribs, where the testicles are felt, prominent at the sub-lumbar region. Then, with the fingers half flexed, the adhesions of the organ are broken off, and the organ, held in the bend of the finger, is brought outside. The second testicle is removed by the same process. If the testicles should slip from the grasp of the finger, the accident is of little account, as they will graft themselves upon the walls of the abdomen, and in time disappear by resolution. The operation is concluded by the closing of the wound by stitches of interrupted suture, and the healing usually takes place by first intention.

The operation is performed in the hen in the same manner as with the male bird, the ovaries being found in the lumbar region, from which they are removed by the same manipulations that are employed in caponing.

Young fowls about in their third month, are usually selected as the victims of this epicurean barbarity.

CONTENTS.

Abdominal Cryptorchidy 77
Abscesses in Scrotal region 89
   "   " Pelvic cavity 130
Advantages of Castration in cows 108
Age at which Castration ought to be performed 8
Amaurosis 103
Anatomy of the male parts 15
   "   " female " 109

Bistournage 58
Bistoury Caché 117

Castration of Cryptorchids 67
   "   Dogs 136
   "   Fashion and convenience 5
   "   Females 106
   "   Fowls 139
   "   Necessity 4
   "   Small ruminants 131
   "   Swine 132
Champignon 90
   "   Superficial 92
   "   Deep 93
Charlier's Instruments 115, 116
   "   Method 115
   "   Scissors 118
Clamps 41, 42
   " House 35
Classification of Methods 19
Colics 85
Combined Forceps of M. Beaufils 30

CONTENTS.

Comparative View of the Various Modes	103
Complications and their Treatment 84,	129
Conditions favorable to the operation	108
Constipation	131
Covered Operation	44
Crushing of the Testicular Cord	57
Curved Clamps	95
Dangers of Operation Standing Up	12
Definition of Castration	3
Dogs	136
Double Subcutaneous Torsion	58
Ecraseurs	33
Effects of Castration 5, 78,	108
Emphysema, subcutaneous	131
Excision	20
Farmer Miles	68
Firing	35
" Forceps for Castration by	69
Fistula of the Scrotum	99
Forceps for Holding the Ovary	118
Fowls	139
Free Torsion	24
Gangrene	88
Hemorrhage 86,	130
Hernia of Castration	99
History of Castration	4
Hygiene and subsequent attention	82
Incision in the Vagina	119
Inguinal Cryptorchidy	75
" Hernia	99
Ligature	51
" Carrier	97
" of the Cord and its envelopes	52

Ligature of the Cord only.................................. 52
" of the Efferent Canal......... 55
" of the Spermatic Artery......................... 54
Limited Torsion...... 26
Linear Crushing........... 32

Method by the Clamps.......... 39
"   " Ecraseur...................... 34
"   " Flanks............................ 110
Methods, Classification of............................ .. 19
Modes of Cicatrization............................ 81
Modus operandi of Bistournage 59
"   "   "   " 1st Step of.............60, 61
"   "   "   " 2d " 62, 63
"   "   "   " 3d " 64, 65
"   "   "   " 4th "   66
"   "   " Castration of Females................ 110
"   "   " Covered Operation....... 44
"   "   " Limited Torsion..................... 28
"   "   " Uncovered Operation...... 47

Peritonitis..100, 130
Preparations of the Patient............................. 9
Purposes of Castration.................................. 4

Removal of the Ovaries ....·............ 120
Restraint, Modes of...................................... 10

Season most favorable........................... 8
Scraping......... 21
Second Method of Castration 39
Securing the Animal Down............................ 10
Softening of the Bags.............. . ...60, 61
Standing Up Operation.................................... 10
"   " Dangers of...................... 12
Subcutaneous Twisting......... 58
Swelling of the Scrotal Region........................... 87
Swine 132

Tearing and Torsion........... 22

CONTENTS.

Tearing of the Clamps.....................................	85
Tetanus..	102
Thimble for Castration.....................................	118
Third Method of Castration.................................	56
Torsion..	23
" above the Epididimis................................	24
" below the Epididimis................................	25
" Forceps...................................26,	27
" Free...	24
" Limited..	26
Uncovered Operation..	47
Vaginal Speculum...................................116,	117

William R. Jenkins's
VETERINARY BOOKS.

850 SIXTH AVENUE, NEW YORK.

Any of the following books will be sent post paid on receipt of the price ; full Catalogue on application.

	PRICE
Animal Castration. By Dr. A. Liautard. 12mo, illustrated	$2 00
American Veterinary Review. Edited by Prof. A. Liautard, H.F.R.C.V.S. Issued monthly. Subscription, $4 per year ; single copy	50
Armatage. "Every Man His Own Horse Doctor." In which is embodied Blaine's "Veterinary Art," with 330 original illustrations, colored plates, anatomical drawings, etc. 8vo, half leather	7 50
Armatage's Veterinarian's Pocket Remembrancer. By George Armatage, M.R.C.V.S., with concise directions and memoranda for the treatment in urgent or rare cases. 32mo, cloth	75
Armatage. Horse owners' and Stable-men's Guide. Crown 8vo, cloth	2 00
Baucher. New Method of Horsemanship. Including the Breaking and Training of Horses. 12mo, cloth, illustrated	1 00
Chauveau. The Comparative Anatomy of the Domesticated Animals. By A. Chauveau, Professor at Lyons Veterinary School, France. New edition, translated, enlarged, and revised. By George Fleming, F.R.C.V.S. 8vo, cloth, with 450 illustrations	6 00

	PRICE
Clok. "The Diseases of Sheep." With proper Remedies to Prevent and Cure the same. By Henry Clok, V.S. 12mo, cloth..	1 25
Clarke. Horse's Teeth. A Treatise on their Mode of Development, Physiological Relations, Anatomy, Pathology, Dentistry, etc. By W. H. Clarke. Revised, enlarged, and illustrated edition. 12mo, cloth.........	2 00
Clater's "Every Man His Own Cattle Doctor." By Francis Clater. New edition, entirely rewritten by George Armatage, with numerous plain and colored plates. 8vo, half leather.............................	7 50
Cobbold. "The Internal Parasites of our Domesticated Animals." A manual of the entozoa of the ox, sheep, dog, horse, pig, and cat. By T. Spencer Cobbold, M.D., F.R.S. 12mo, cloth, illustrated.....................	2 00
Dalziel. "British Dogs." Their Varieties, History, Characteristics, Breeding, Management, and Exhibition. Illustrated with full page portraits. 12mo, cloth........	4 00
Dalziel. Diseases of Dogs. 12mo, cloth..................	1 00
Dana. "Tables in Comparative Physiology." Giving Comparative Weight, Temperature, Circulation of the Blood, Respiration, Digestion, Nervous Force and Action between Man and the Lower Animals and Birds. By Prof. C. L. Dana, M.D. Chart on paper...	25
Day. The Race horse in Training. With some hints on Racing and Racing Reform. By Wm. Day. Demy 8vo..	6 40
Dun. Veterinary Medicines. Their Actions and Uses. By Finlay Dun, V.S. New American edition from the latest English one. 8vo, cloth......................	3 50
New Revised English edition, 8vo, cloth..............	5 00
Fearnley. Lessons in Horse Judging, and on the Summering of Hunters. 12mo, cloth, illustrated.........	1 60
Fearnley. Lecture on the Examination of Horses as to Soundness, Sale, and Warranty. By W. Fearnley, M.R.C.V.S. 12mo, cloth............................	3 00
Fitzwygram. Horses and Stables. By Col. F. Fitzwygram of the 15th Hussars. New edition. With 24 illustrations. Cloth.............................	4 00
Fleming. "Human and Animal Variolæ." A study in Comparative Pathology. Paper.....................	25
Fleming. "The Contagious Diseases of Animals." Their Influence on the Wealth and Health of Nations. 12mo, paper...	25

W. R. Jenkins's List of Veterinary Books. 3

PRICE

Fleming. "Actinomykosis." A New Infectious Disease of Man and Animals. By George Fleming, F.R.C.V.S. Paper. (Just published.)............................ 25

Fleming. On Horseshoeing. By Geo. Fleming. Cloth... 75

Fleming. Operative Veterinary Surgery. By George Fleming. (In Preparation.) Part 1 now ready.

Fleming. Propagation of Tuberculosis. By George Fleming. Cloth... 2 25

Fleming. Manual of Veterinary Science and Sanitary Police. Embracing the Nature, Causes, Symptoms, etc., and the Prevention, Suppression, Therapeutic Treatment, and the Relation to the Public Health of the Epizootic and Contagious Diseases of the Domesticated Animals; with a scheme for Veterinary and Sanitary Organization; Observations on the Duties of Veterinary Inspectors, Legislative Measures, Inspection of Meat and Milk, Slaughter Houses, etc. By George Fleming, F.R.G.S. 2 vols., 8vo, cloth, illustrated...... 9 00

Fleming. "Animal Plagues." Their History, Nature, and Prevention. By George Fleming, F.R.C.V.S., etc. Being a Chronological History from the earliest times to 1844. First Series, comprising a History of Animal Plagues from B.C. 1490 to A.D. 1800. 8vo, cloth...... 6 00
Second Series, containing the History from A. D. 1800 to 1844. 8vo, cloth, (recently published)............ 4 80

Fleming. Veterinary Obstetrics. Including the Accidents and Diseases incident to Pregnancy, Parturition, and the Early Age in Domesticated Animals. By Geo. Fleming, F.R.C.V.S. With 212 illustrations. 8vo, cl. 6 00

Fleming's Rabies and Hydrophobia. History, Natural Causes, Symptoms, and Prevention. By George Fleming, M.R.C.V.S. 8vo, cloth..................... 6 00

Going. Veterinary Dictionary. Compiled by Prof. J.A.Going. 2 00

Hayes. Veterinary Notes for Horse Owners. An Everyday Horse Book. Revised edition, illustrated. By M. H. Hayes. 12mo, cloth............................$5 00

Heatley. The Horse-owners' Safeguard. A handy Medical Guide for every Horse-owner. By George S. Heatley, V.S. 12mo, cloth..................................... 2 00

Hill. The Management and Diseases of the Dog. Containing full instructions for Breeding, Rearing, and Kennelling Dogs. Their different Diseases, embracing Distemper, Mouth, Teeth, Tongue, Gullet, Respiratory Organs, Hepatitis, Indigestions, Gastritis, St. Vitus' Dance, Bowel Diseases, Paralysis, Rheumatism, Fits,

4   W. R. Jenkins's List of Veterinary Books.

PRICE

Rabies, Skin Diseases, Canker, Diseases of the Limbs, Fractures, Operations, etc. How to detect and how to cure them. Their Medicines, and the Doses in which they can be safely administered. By J. Woodroffe Hill, F.R.C.V.S. 12mo, cloth extra, fully illustrated.. 2 00

Hill. "The Principles and Practice of Bovine Medicine and Surgery." By J. Woodroffe Hill, F.R.C.V.S. Octavo, 664 pages, with 153 illustrations on wood and 19 full page colored plates. Cloth..................... 10 00
octavo, 664 pages, sheep............ .... 11 50

Holcombe. "Laminitis." A Contribution to Veterinary Pathology. By A. A. Holcombe, V.S. Pamphlet..... 50

Horses and Roads ; or, How to Keep a Horse Sound on his Legs. By "Free Lance"................ .... 2 50

Howden. "How to Buy and Sell the Horse." The object of this book is to explain in the simplest manner what constitutes a sound horse from an unsound one. 12mo, cloth......................... .... 1 00

Jennings. Horse Training Made Easy. A Practical System of Educating the Horse. By Robert Jennings, V.S. 12mo, cloth............ .... 1 25

Jennings. Swine, Sheep, and Poultry. Embracing a History and Varieties of each; Breeding, Management, Disease, etc. By Robert Jennings, V.S. 12mo, cloth. 1 25

Jennings. Cattle and their Diseases; with the best Remedies adapted to their Cure. By Robert Jennings, V.S. 12mo, cloth......................... .... 1 25

Jennings. On the Horse and his Diseases. By Robert Jennings, V.S. 12mo, cloth..................... 1 25

Journal of Comparative Medicine and Surgery. A Quarterly Journal devoted to the Diseases of Animals, particularly of the Horse. Published in January, April, July, and October. Subscriptions, $2 per annum. Single copies, postpaid......................... 60

Laverack. The Setter. By E Laverack. With instructions how to Breed, Rear, Break, etc. Colored illustrations............. .... 3 00

Liautard. Vade Mecum of Equine Anatomy. By A. Liautard, M.D., V.S. 12mo, cloth........................ 1 75

Liautard. "Animal Castration." By Dr. A. Liautard, D.V.S. 12mo, illus. $2 00

Liautard. Translation of Zundel on the Horse's Foot. By Dr. A. Liautard, D.V.S. 8vo, cloth................... 1 50

Law. The Lung Plague of Cattle ; Contagious Pleuro-Pneu-

W. R. Jenkins's List of Veterinary Books. 5

	PRICE
monia. Illustrated. By James Law, Professor of Veterinary Medicine in Cornell University. Paper, 100 pp.	30
Law. Farmers' Veterinary Adviser. A Guide to the Prevention and Treatment of Disease in Domestic Animals. By James Law, Professor of Veterinary Medicine in Cornell University. Illustrated. 8vo, cloth..	3 00
Lehndorff. Horsebreeding Recollections. By G. Lehndorff. 8vo, cloth.	4 20
Martin. Cattle. Their Various Breeds, Management, and Diseases. By W. C. L. Martin. Revised by W. Raynbird. 16mo, boards.	50
McAlpine. Biological Atlas. Containing 24 plates of 423 colored illustrations. Oblong quarto cloth. By D. McAlpine, F.C.S.	3 00
McBride. Anatomical Outlines of the Horse. Revised and Enlarged by T. M. Mayer, M.R.C.V.S. With colored illustrations. 12mo, cloth.	3 40
McClure. Diseases of American Horses, Cattle, and Sheep. Their Treatment; with full description of the Medicines employed. By R. McClure, M.D., V.S. 12mo, cl., illus.	2 00
McClure. American Gentlemen's Stable Guide; with the most Approved Methods of Feeding, Grooming, and Managing the Horse. By Robert McClure, M.D., V.S. 12mo, cloth.	1 00
Meyrick. Stable Management and the Prevention of Diseases among Horses in India. By J. J. Meyrick, F.R.C.V.S. 12mo, cloth.	1 00
Miles. Remarks on Horses' Teeth. Addressed to Purchasers. By W. Miles.	60
Moreton. "On Horsebreaking." By Robert Moreton. 12mo, cloth.	50
Moreton's Manual of Pharmacy for the Veterinary Student. By J. W. Morton. 12mo, cloth.	4 00
Navin. "The Explanatory Stock Doctor," for the use of the Farmer, Breeder, and Owner of the Horse. With numerous illustrations. By John Nicholson Navin, V.S. 8vo, sheep.	4 75
Percival. Hyppo-pathology. A Systematic Treatise on the Disorders and Lameness of the Horse. By W. Percival. With many illustrations. 6 vols., boards.	34 20
Percival. Lectures on Horses; Their Form and Action. By W. Percival. With eight outline plates. 8vo, cloth.	4 00
Percival's Anatomy of the Horse. By W. Percival. 8vo, cloth.	8 00

	PRICE
Peck. "Classifications of the Muscles of the Horse." This is a large chart, printed on heavy paper, 24x38 inches, showing at a glance the Classifications of the Muscles of the Horse, with Origin, Insertion, Nervous Supply, and the Function of each....................................$	50
Reynolds. "Breeding and Management of Draught Horses." By Richard S. Reynolds, M.R.C.V.S. Crown 8vo, cl....	1 40
Riley. The Mule. A Treatise on the Breeding, Training, and Uses to which he may be put. 12mo, cloth, illus..	1 50
Robertson. The Practice of Equine Medicine. By W. Robertson..	6 00
Steel. A Treatise on the Diseases of the Ox. Being a Manual of Bovine Pathology, especially adapted to Veterinary Practitioners and Students. By John Henry Steel, M.R.C.V.S., F.Z.S. 8vo, with 118 illus., cl......	6 00
Steel. "Outlines of Equine Anatomy." A Manual for the use of Veterinary Students in the Dissecting Room. By John H. Steel, M.R.C.V.S. 12mo, cloth...........	3 00
Strangeway. "Veterinary Anatomy." New edition, revised and edited by I. Vaughn, F.L.S., M.R.C.V.S., with several hundred illustrations. 8vo, cloth............	8 00
Stornmouth's Manual of Scientific Terms. Especially referring to those in Botany, Natural History, Medical and Veterinary Science. By Rev. Jas. Stornmouth....	3 00
Tellor. "Diseases of Live Stock," and their most Efficient Remedies. By Lloyd V. Tellor. 8vo, cloth, illustrated, $2.50; sheep..	3 00
Tuson. Pharmacopœia, including Outlines of Materia Medica and Therapeutics in Veterinary Medicine. By R. V. Tuson. 12mo, cloth.........................	2 50
Veterinary Diagrams. Five Charts, each 22x28 inches in size, on stout paper, as follows, sold separately:	
No. 1, with eight colored illustrations. External Form and Elementary Anatomy of the Horse...............	1 50
No. 2. Unsoundness and Defects of the Horse, with 50 woodcuts..	75
No. 3. The Age of the Domestic Animals, with 42 woodcuts..	75
No. 4. The Shoeing of the Horse, Mule, and Ox, with 59 woodcuts...	75
No. 5. The Elementary Anatomy, Points, and Butcher's Joints of the Ox, with 17 colored illustrations. With explanatory text.......................................	1 50
Price per set of five.................................	5 00

W. R. Jenkins's List of Veterinary Books. 7

PRICE

Walley. "Four Bovine Scourges." (Pleuro-Pneumonia, Foot and Mouth Disease, Cattle Plague, and Tubercle.) With an Appendix on the Inspection of Live Animals and Meat. By Thos. Walley, M.R.C.V.S. With 49 colored illus. and numerous woodcuts. 4to, cl.........$6 40

Webb. "On the Dog." Its Points, Peculiarities, Instsinct, and Whims. Illustrated with photographs............ 3 00

Williams. Principles and Practice of Veterinary Medicine. New edition, entirely revised. and illustrated with numerous plain and colored plates. By W. Williams, M.R.C.V.S. 8vo, cloth.............................. 5 00

Williams. Principles and Practice of Veterinary Surgery. New edition, entirely revised, and illustrated with numerous plain and colored plates. By W. Williams, M.R.C.V.S. 8vo, cloth.............................. 7 50

Williams. Chart of the Contagious, Infectious, and Specific Fevers of the Domesticated Animals.................. 1 00

Zundel. "On the Horse's Foot." Translated by A. Liautard, M.D., D.V.S....................................... 1 50

VETERINARY BOOKS IN FRENCH.

Benion. Traité de l'Élevage et des Maladies des Animaux et des Oiseaux de Basse-Cour.......................$2 80

Benion. Traité de l'Élevage et des Maladies du Mouton.. 3 60

Benion. Traité de l'Elevage et des Maladies du Porc..... 2 60

Beugnot. Dictionnaire usuel de Chirurgie et de Médecine Vétérinaire. 2 forts volumes in-8, avec planches....... 7 20

Bouley. La Rage, moyen d'en éviter les Dangers et de prévenir sa Propagation............................ 40

Bouley-Reynal. Nouveau Dictionnaire Pratique de Médecine, de Chirurgie et Hygiène Vétérinaire (to be completed in 18 volumes), chaque volume................ 3 00

Colin. Traité de Physiologie Comparée des Animaux; Par G. Colin, Professeur à l'école Vétérinaire d'Alfort; avec Figures intercalées dans le texte. 2 vols. in-8....10 40

Cruzel. Des Maladies de l'Espèce Bovine. Par J. Cruzel. 5 60

Dictionnaire. Lexicographique et Descriptif des Sciences Médicales et Vétérinaires. Un très-fort vol. de plus de 1500 pages.. 8 00

Gourdon. Traité de la Castration des Animaux Domestiques 3 60

	PRICE
Hertwig. Les Maladies des Chiens et leur Traitement...	$1 40
Lecocq. Traité de l'Extérieur du Cheval et des Princip aux Animaux Domestiques................	3 60
Leyh. Anatomie des Animaux Domestiques.............	3 60
Magne. Races Chevalines et leur Amélioration, Entretien, Multiplication, Élevage et Éducation du Cheval, de l'Ane et du Mulet. Par J H. Magne.................	3 20
Magne. Races Bovines et leur Amélioration, Entretien, Multiplication, Élevage et Engraissement du Bœuf. Par J. A. Magne	2 00
Magne. Races Porcines et leur Amélioration, Entretien, Multiplication, Élevage et Engraissement du Porc. Par J. H. Magne.......................................	80
Magne. Nourriture des Chevaux de Travail—brochure.....	40
Magne. Choix du Cheval................................	80
Magne. Choix et Nourriture du Cheval. Par J. H. Magne. Avec Vignettes..	1 40
Mourod. Matière Médicale ; ou la Pharmacologie Vétérinaric.	2 40
Saint-Cyr. Traité d' Obstétrique Vétérinaire. Avec cent vignettes	5 60
Signol. Aide Mémoire du Vétérinaire, Médecine, Chirurgie et Obstétrique ; Par Jules Signol ; avec 395 Figures...	2 40
Tabourin. Nouveau Traité de Matière Médicale Thérapeutique et de Pharmacie Vétérinaires, 2 fort volumes, in-8, avec plus de 100 figures.	8 60

www.ingramcontent.com/pod-product-compliance
Lightning Source LLC
Chambersburg PA
CBHW030254170426

43202CB00009B/745